SOCIETY AND CLIMATE

Transformations and Challenges

SOCIETY AND CLIMATE
Transformations and Challenges

Nico Stehr
Zeppelin University, Germany

Amanda Machin
Zeppelin University, Germany

World Scientific

NEW JERSEY · LONDON · SINGAPORE · BEIJING · SHANGHAI · HONG KONG · TAIPEI · CHENNAI · TOKYO

Published by

World Scientific Publishing Co. Pte. Ltd.
5 Toh Tuck Link, Singapore 596224
USA office: 27 Warren Street, Suite 401-402, Hackensack, NJ 07601
UK office: 57 Shelton Street, Covent Garden, London WC2H 9HE

Library of Congress Control Number: 2019943487

British Library Cataloguing-in-Publication Data
A catalogue record for this book is available from the British Library.

SOCIETY AND CLIMATE
Transformations and Challenges

ISBN 978-981-3272-42-2

For any available supplementary material, please visit
https://www.worldscientific.com/worldscibooks/10.1142/11047#t=suppl

Desk Editor: Amanda Yun

Typeset by Stallion Press
Email: enquiries@stallionpress.com

Printed in Singapore

This book is dedicated to our children and our children's children.

Acknowledgements

This book draws upon previous material written by Hans von Storch and Nico Stehr. We would like, first and foremost, to thank Hans von Storch for his valued ongoing support. We would also like to thank Keigo Akimoto, Jun Arima, Roger A. Pielke, Sr., Alexander Ruser, Franz Mauelshagen and Thomas Bruhn for helpful suggestions on the manuscript, Oliver Wagener and Falk Stratenwerth for their research and editorial assistance and Beate Gardeike for preparing and enhancing the graphical material. We would like to express our deep gratitude to Volker and Roswitha Heuer, without whose support this project would not have been possible. The Institute of Advanced Sustainability Studies (IASS) in Potsdam, and Zeppelin University in Friedrichshafen, have both provided the opportunities for inspiring discussion and contemplation on the topic of climate and society in general.

Contents

Overview

This book starts from the premise that the interconnection between climate and society is a fascinating and fertile topic for analysis. The climate might once have provided a taken-for-granted background against which social, political and economic interactions could take place. If climate conditions stay the same from year to year, if the seasons unfold as expected, the future can be envisioned and planned; architectural designs, agricultural rotations and travel schedules can be constructed as firm fixtures within a world that is, in so many other ways, in a state of flux. But this taken-for-granted background is cleaving. Climate issues have migrated from the background to the forefront of concerns across society, appearing in newspaper headlines, political party manifestos and everyday conversations, as well as in academic research of various disciplines. Even if one contests dominant depictions and discourses, they are nonetheless becoming increasingly hard to ignore. *Climate matters*, and it matters in ways that we are only just beginning to understand.

But the issue is not just that a volatile climate will impact society. For what has become hard to deny, although this has not meant that it is entirely accepted, is the way that social institutions, and the energy base that underpins them, affects the climate. So just as climate matters to society, *society matters* to the climate (Urry, 2011: 3). Industrial economies, totally dependent upon fossil fuels, have released so much greenhouse gas into the atmosphere that climate will be affected irrevocably. Although commonly referred to as "global warming", the expected consequences are not only increasing average global temperatures, but rising sea levels and more frequent occurrences of extreme weather. This in turn will have serious consequences for society for a long time to come. This is why German Chancellor Angela Merkel is convinced that anthropogenic climate change is "one of the greatest challenges facing humanity" (Osborne, 2017). Climate change is, for many, understood to be a catastrophe of unique proportions, something that, in the words of Naomi Klein (2014) "changes everything".

Of course, *framing* climate change as a challenge within political rhetoric does not necessarily underpin a robust political response adequate to propel the transformation demanded by Klein. The capacity and willingness of contemporary societies to respond to the challenges of climate change is somewhat unclear to say the least. Amitav Ghosh diagnoses a pervasive cultural complicity in the blindness to the threats of a changing climate, what he calls "the great derangement" (Ghosh, 2016). So while the issue of global warming has moved to the foreground of global politics, the enthusiastic rhetoric may belie a more ambiguous reality. The Paris climate treaty, for example, is an accord within the United Nations Framework Convention on Climate Change. Adopted by consensus on 12 December 2015 by 195 countries, it deals with greenhouse gas mitigation, adaptation and finance commencing in 2020. Its stated first aim is "holding the increase in the global temperature to well below 2°C above pre-industrial levels and to pursue efforts to limit the temperature increase to 1.5°C above pre-industrial levels, recognising that this would significantly reduce the risks and impacts of climate change" (United Nations, 2015: 22). Widely regarded as an historical achievement (see Milman *et al.* (2017)), the Paris treaty seemingly marks a general scientific and public consensus that anthropogenic climate change is a very serious threat.

The treaty, however, is not legally binding. There are no formal sanctions if a country should fail to live up to its commitments regarding efforts in terms of mitigation, adaptation or finance, and there is no guarantee how far-reaching the Paris agreement will be. This problem came to the fore on 1 June 2017 when the President of the United States, Donald Trump, announced its formal withdrawal from the treaty. At the time, the United States joined Nicaragua and Syria as the only countries in the world outside of the climate treaty. Nicaragua and Syria later signed. Although under the terms of the Paris agreement, the United States cannot formally begin the process of withdrawal until November 2019, the current administration is already embarked upon a strong anti-environmentalist agenda. In his announcement,

Trump was fulfilling his campaign pledge to "end the war on coal" (Bomberg, 2017: 1) and his purported aim to reclaim sovereignty for the American people and put "America first" (Milman *et al.*, 2017). Shortly after his announcement in which Trump emphasised that he was elected to represent "the people of Pittsburgh and not Paris", the mayor of the city of Pittsburgh, Bill Peduto, voiced his criticism of the withdrawal and the city's commitment to the treaty (see Gambino (2017)). Indeed, a number of American states and cities will continue to follow its announced climate policies, offering "a profound counter to Trump's anti-environmental crusade" (Bomberg, 2017: 5; Cooper and Ronayne, 2017). What this scenario illustrates, at the very least, is the high degree of politicisation of the issues of climate and climate change in the contemporary world.

By pulling out of the agreement, many regard the United States (currently the second largest carbon dioxide emitter, after China) as failing to fulfil its ethical obligations regarding climate change. Those parts of the world that will be most existentially challenged by anthropogenic climate change are those who benefitted the least from the industrialisation that instigated its onset. Concomitantly, they are both more vulnerable to its risks and less well-resourced to negotiate for justice (Roberts and Parks, 2007). The exacerbation by climate change of already existing global inequalities reveals climate to be a highly complex system interlacing geophysical elements with social, economic and political processes and institutions.

We draw to attention three inadequacies that characterise common accounts of climate — accounts that understand it as a *global* object of *natural science* and of only relatively *recent* interest. First, climate is predominantly depicted at a *global* level. And yet the impact of global environmental problems as well as their causes are unequally distributed around the globe. To describe climate change as a global issue, imparts its seriousness, but belies its differentiated causes and consequences. It is not human society *in general* that has heightened the Greenhouse Effect through industrialisation, urbanisation and deforestation. It is not humanity as a homogenous whole who will suffer the brunt of the repercussions. And yet the locally differentiated impact is precisely a result of the way in which climate works at a planetary level, as part of a sensitive and complex system, distributing effects at a different scale to their causes.

Secondly, climate is frequently understood to be an issue of *natural science* (see Grundmann and Stehr (2000; 2010)). But climatic concerns cannot be separated out of their social context and therefore fall into the remit of social science too. The climate changes not only because of natural processes, but also because of social technologies, economic patterns and cultural behaviour. Conversely, social institutions and political relations may be affected by climate and its vicissitudes; inequalities of gender, race and class may be sharpened; discourses of security and economic growth may be reformulated to incorporate environmental risk. Climate and society are enfolded together. This means that analysis is needed not only by geophysicists, biogeochemists and meteorologists, but also by sociologists, philosophers, historians, anthropologists and political scientists. At the same time, it is also important to consider lay perceptions of climate. For without public support, the success of any political movement or policy decision on climate is unlikely. But just as the input of scientific knowledge is crucial for public understanding on this issue, the tension between scientific expertise and public perception of climate is becoming evident. Different constructions of climate condition, complement and contradict each other.

Third, because of the recent prominence of anthropogenic climate change, climate has been widely supposed to be a *recent* topic of analysis. But climate and its variability and its implications for society have long been investigated; colonialists and chemists, mountaineers and mariners, have attended to the various components of the climate in relation to an array of different agendas and assumptions. These previous accounts are worth investigating, and not just for the sake of academic curiosity. Their discoveries, ideas and mistakes may reveal something about our own, and may dissuade us from proclaiming novelties when there are none, and from retracing erroneous steps that have been trod before.

It is therefore crucial to put recent research on climate into historical perspective. Contemporary warnings of a changing climate are, for sure, urgent and alarming. Notwithstanding the uniqueness of the current climate situation, there may be something to gain from considering such warnings in the context of historical accounts of climate and climate change. (Re)familiarising ourselves with previous scientific research should not be misunderstood as an attempt to preserve out-of-date ideas, but can be helpful in the production of new ones (Stehr and von Storch, 2000: 2).

Recognition that grasping the scope, meaning and implications of climate and climate change requires input from different disciplines engaging with various geographical and temporal scales, opens up the field of enquiry, provoking numerous questions: How "natural" is climate? How "unnatural" is climate change? What are the social, political and cultural implications of a changing climate? Does it finally reveal the ultimate fragility of social life or rather its creative power? Can climate be controlled by human technology or is this a hubristic assumption that belies the profound complexity of the climate system? How do scientific and "everyday" renderings of climate and climate change differ? What might be learnt from placing our current scientific and common-sense constructions next to previous meanings? And what can be learnt from social responses to environmental concerns, past and present?

It is such potent queries regarding the interconnections between human beings, societies, sciences, natures and climates that we hope to draw attention to in this book, although we do not pretend to be able to provide solid answers. Our main intention is to offer a broad picture of the perplexing but revealing interconnections between climate and society and to expand the horizon of analysis. We do not intend to imply that this picture is a comprehensive one, conscious as we are of our position as social science scholars of the global north and of the many lacunae and biases in our account.

Chapter One provides an introduction to the changing approach to climate, by considering it in terms of the dichotomy of "nature" and "society". We point out that diverse meanings of "climate" are found across different periods of history, different parts of the globe and across the manifold fields of social activity; religious, scientific and everyday understandings complement and compete. Climate emerged as a scientific object around the eighteenth century. But we cannot see this object as separate from the social structures it both delimits and is affected by. Climate change can be understood as one facet of the "Anthropocene", a geological epoch attesting to the impact of certain form of human societies on the planet. With the Anthropocene, the very distinction between nature and society collapses, and climate becomes both natural *and* social.

Chapter Two is concerned with the emergence of climate as a scientific object. We start by attending to the scientific rendering of climate as *average weather*. We see how this rendering produced an idea of a set of regular conditions that can be utilised as a reliable resource. We describe a shift in climatology, from a generator of statistical and numerical accounts of regional climate in the nineteenth century, towards a global analytical approach, when climate became a complex structured system in which the components of atmosphere, hydrosphere, cryosphere and biosphere are enfolded together. Climate has become *possible weather*.

Scientists today are far more likely to highlight the *instability* of climate over its reliability. Chapter Three, then, focuses upon this dimension of climate: climate as a variable set of conditions. Here we turn to the issue of climate *change* and consider how it was considered in the past and today. While climate variation has long been a topic of research, it was understood previously to occur periodically, or in a cycle. The difference today is that the changes to the climate that are seen to be important are those that are "progressive" and thus irreversible. We also consider the unequally distributed impacts of climate change, noting that the effects of "natural" environmental hazards and risks cannot be separated out from particular socio-political conditions, such as power relations, gender norms and social inequalities.

What is the extent of the impact of climate change on society? Chapter Four delves into the murky past of climate determinism and considers the claim that the possibilities for human civilisation are ultimately decided by climate. This idea was understood to be a rigorous scientific fact in the nineteenth century. We warn that there may be a return to climate determinism in current analysis and that this should be carefully monitored. For, whereas climate is a *condition* of human society, it is not a *determinant*, and to forget this is to undermine the power of human and social agency. However, this by no means implies that climate has no influence whatsoever. Indeed, the disruptions and dangers of a changing climate in the coming decades may well have enormous impact upon societies. We can neither control the climate nor insulate ourselves from its volatility. This has perhaps never been as clear as today, when the potentially dramatic repercussions of a changing climate are at the forefront of social, political and economic discussion.

In Chapter 5 we turn to more general public perceptions of climate and the tendency to assume a set of stable climate conditions that can be trusted to remain constant. Paradoxically, such trust may serve to blind us to the ways in which societies are working to undermine those very conditions. Indeed, the potentially dramatic consequences of anthropogenic climate change are becoming hard to ignore. Alongside the depiction of climate as constant condition is a newly dominant depiction of climate as a source of future catastrophe. Ostensibly contradictory, we consider whether both these depictions of climate — as constant and as catastrophe — both preclude a lack of social and political engagement. What role do climate scientists play here? We also consider the issue of (mis)trust of climate scientists in the public sphere. What we hope to highlight in this chapter is the way in which perceptions of climate are deeply entrenched in our forms of life and cannot be easily displaced. This has immense implications for policymaking.

Chapter 6, then, presents climate as a policy issue. In line with the assertion we make above, that climate is no longer simply just background condition of everyday life, we notice how climate has emerged as a policy area in its own right, one that demands active assessment and response. The changing climate along with its concomitant consequences and risks, is now incorporated into our political institutions and mechanisms as an important issue for which policy should be determined and implemented. We distinguish three distinct, but not mutually exclusive, approaches to climate policy: technological, economic and regulatory.

Climate has not only moved to the centre stage in the political arena, but in the academic arena too. In our short conclusion, we consider the implications of this growing prominence of climate for the academy. We suggest that climate change opens the possibilities for new types of interdisciplinary work. Despite wariness around scientific contributions to public policy, new ways of undertaking science, conceptualising science and applying science are urgently needed if climate change is going to be tackled effectively, creatively and fairly. Is a new phase in understanding the connection between climate and society on the horizon?

References

Bomberg, Elizabeth (2017) "Environmental politics in the Trump era: An early assessment," *Environmental Politics* **26**(5): 956–963.

Cooper, Jonathan J. and Kathleen Ronayne (2017) "Proposed California climate deal takes aim at toxic air," *Associated Press*. 12 July. Available at: www.apnews.com/63ba378c79714172a54206882d9ca160/Proposed-California-climate-deal-takes-aim-at-toxic-air (accessed: 14 March 2018)

Gambino, Lauren (2017) "Pittsburgh fires back at Trump: We stand with Paris, not you," *The Guardian*. 12 June. Available at: www.theguardian.com/us-news/2017/jun/01/pittsburgh-fires-back-trump-paris-agreement (accessed: 14 March 2018).

Ghosh, Amitav (2016) *The Great Derangement: Climate Change and the Unthinkable*. London and Chicago: University of Chicago Press.

Grundmann, Reiner and Nico Stehr (2000) "Social science and the absence of nature," *Social Science Information* **39**(1): 155–179.

Grundmann, Reiner and Nico Stehr (2010) "Climate change: What role for sociology?" *Current Sociology* **58**(6): 897–910.

Klein, Naomi (2014) *This Changes Everything: Capitalism vs. the Climate*. New York: Simon and Schuster.

Milman, Oliver, David Smith and Damian Carrington (2017) "Donald Trump confirms US will quit Paris climate agreement," *The Guardian*. 1 June.

Available at: www.theguardian.com/environment/2017/jun/01/donald-trump-confirms-us-will-quit-paris-climate-deal (accessed: 14 March 2018)

Osborne, Samuel (2017) "Angela Merkel promises to tackle Donald Trump on climate change at G20 summit," *The Independent*. 29 June. Available at: www.independent.co.uk/news/world/europe/angela-merkel-donald-trump-g20-summit-climate-change-germany-chancellor-us-president-trade-wilbur-a7813716.html (accessed: 14 March 2018)

Roberts, J. Timmons and Bradley C. Parks (2007) *A Climate of Injustice: Global Inequality, North-South Politics, and Climate Policy*. Cambridge, London: MIT Press.

Stehr, Nico and Hans von Storch (2000) "Eduard Brückner's ideas — Relevant in his time and today". In: *Eduard Brückner — The Sources and Consequences of Climate Change and Climate Variability in Historical Times*. (Eds.) Stehr, Nico and Hans von Storch. Springer.

United Nations (2015) Adoption of the Paris Agreement. Available at https://unfccc.int/resource/docs/2015/cop21/eng/l09r01.pdf

Urry, John (2011) *Climate Change and Society*. Cambridge and Malden: Polity Press.

CHAPTER 1

Introduction: Society, Nature, Climate

The term climate, taken in its most general sense, indicates all the changes in the atmosphere, which sensibly affect our organs, as temperature, humidity, variations in the barometrical pressure, the calm state of the air or the action of opposite winds, the amount of electric tension, the purity of the atmosphere or its admixture with more or less noxious gaseous exhalations, and finally, the degree of ordinary transparency and clearness of the sky, which is not only important with respect to the increased radiation from the earth, the organic development of plants, and the ripening of fruits, but also with reference to its influence on the feelings and mental condition of men.

Alexander von Humboldt (1856: 317)

Climate constitutes one of the most important general conditions for human existence. It is no surprise that over the centuries, climate has been a recurrent theme of social, cultural and scientific reflection. Numerous accounts identified climate as an explanatory factor of the nature of human civilisation and its particular forms, successes and failures. In the first volume of his famous treatise *Cosmos, A Physical Description of the Universe*, the German polymath Alexander von Humboldt's broad definition of climate (above) refers to its profound effects on the human condition. The horizon of social expectation and imagination was understood to be set, at least in part, by the climate. Thus, as Eva Horn notices, climate has long been understood as the "imprint of nature upon man" (Horn, 2016: 237).

This relation does not work one way, as "nature" itself bears the distinctive imprint of human societies; forms of agriculture, architecture, transport and technology of different shapes and sizes cultivate and contaminate their environments. This, too, did not go unnoticed by Humboldt, who warned that "the wants and restless activity of large communities of men gradually despoil the face of the earth" (Humboldt, 1849: 11). Today, arguably more than ever, the "imprint of man upon nature" is evident in recent pronouncements of anthropogenic climate change. Along with other environmental issues such as biodiversity loss, disturbance of the phosphorus and nitrogen cycles, acid rain and ozone layer depletion, climate change draws attention to the interconnections and tensions between "nature" and "society" (Latour, 2017). Not only does it highlight human dependency upon clean and stable environmental conditions, it also implicates human societies in the pollution and degradation that threaten clean and stable environmental conditions.

These interconnections between nature and society are not at all uniform. Perceptions of climate and the implications of climate change will therefore vary according to cultural traditions and socio-economic structures (Meinert and Leggewie, 2013). Our starting point in this book is the recognition that diverse meanings of "climate" and its component factors are found through various historical epochs, regions of the world, and across the manifold fields of social activity. We begin by discussing the way in which weather patterns and extremes have been, and continue to be, understood in different ways, long before "climate" emerged as a scientific object.

1.1 The Nature of Weather

Human societies have always been greatly dependent on their surrounding habitats and are deeply affected by environmental conditions and crises, see Stehr (1978). But they have interpreted these conditions and crises in different ways. The weather and its extremes have frequently been given magical or religious interpretation. As James Frazer documents, different rituals and symbolic actions and objects (including fire, frogs and foreskins) were designed to influence the onset of rain or to turn away storms (Frazer, 2002). Magical manipulation of the weather was seen in medieval Iceland, for example, as entirely "natural" and "down to earth" as weather forecasting is today (Ogilvie and Pálsson, 2003). See also Kvideland and Sehmsdorf (1991). Wolfgang Behringer (1999) describes how, from late fourteenth and fifteenth centuries, the existence of "unnatural" climatic phenomena, in particular the onset of the Little Ice Age, were explained as the malign intent of "a great conspiracy of witches". Women denounced as weather witches were burnt at the stake. For Emily Oster this took place as early as the mid-thirteenth century: "In a time period when the reasons for changes in weather were largely a mystery, people would have searched for a scapegoat in the face of deadly changes in weather patterns. "Witches" became target for blame because there was an existing cultural framework that both allowed their persecution and suggested that they could control the weather." (Oster, 2004: 216)

Weather extremes such as floods, storms, droughts and earthquakes and their related phenomena such as plagues, pestilence, cattle epidemics and bad harvests, have also been interpreted as signifiers of the wrath of the gods; punishment for man's sinful conduct (Stehr and von Storch, 2000). Take, for example the case of the "Great European Famine" that occurred between 1315 and 1322 (Lucas, 1930; Jordan, 1996: 7). As Bruce Campbell explains, the famine occurred at a time of heightened climate instability, which involved a "cocktail of natural disasters" (Campbell, 2016: 226). The failure of harvests caused the price of grain and other foodstuffs to rise and a general shortage of resources. The resulting hunger was severe and widespread. Sources suggest that the poor ate "dogs, cats, the dung of doves, and even their own children" (Lucas, 1930: 355). The cause was the incessant rainfall during the summer months, compounded by war (Jordan, 1996: 21). In particular, the torrential rain in 1315 was so unusual that it seemed to be the stuff of biblical prophecy (Lucas, 1930: 346). The foul weather was proclaimed by the churches to be God's punishment: "Sin, hatred of the visible Church, empty faith, and lack of loyalty offended God and explained why He permitted the foul weather to linger for such a length of time" (Jordan, 1996: 22). Prayers, fasting and barefoot processions were held, in order to petition God to "reestablish the normal rhythms of weather" (Jordan, 1996: 108).

Climate disasters can be interpreted as *founding* events. For example, the Great Flood of Gun-Yu, is understood to be fundamental for Chinese culture and the establishment of the Xia dynasty in 1900 BC (Wu et al., 2016). Whether or not there is empirical truth in the story, the myth tells the tale of "the creation of an ordered world out of chaos" (Lewis, 2006: 1). Several versions exist, but many understand the flood as sanctioning political authority and as the origin of Chinese culture (Lewis, 2006: 17). While some versions of the story centre on the emergence of the dynastic state, others construct the flood as punishment and the expelling of criminals.

As Susan Neiman describes, the devastating earthquake that hit Lisbon in 1755 was also interpreted not as a *natural* evil but rather as the result of human "greed and licentiousness" (Neiman, 2015: 243). However, this event of massive material destruction led to conceptual devastation too (Neiman, 2015: 240), precipitating a questioning of God's benevolence and existence by Enlightenment philosophers. In Neiman's words: "At one particular moment in Europe … an earthquake could shake the foundations of faith and call the goodness of Creation into question." (Neiman, 2015: 246). The earthquake came at a time when rationalism was beginning to demand that nature was ultimately knowable: "Natural sciences had combined to confirm Enlightenment conviction that the universe is, as a whole, intelligible." (Neiman, 2015: 246)

Indeed, the modern rationalist hope has not only been for humans to know nature but to be able to master nature, and to reduce the risks posed by volatile weather (Leiss, 1972). This hope, actually, is not restricted to the modern era. Anthropologists discuss the emergence of various strategies (such as storage and cultivation of animals) and technologies (such as fishhooks and nets) aimed at reducing risk, both ecological and social (Hayden, 2009). This altered the relationship between the weather and society; in some ways making human communities less immediately affected by extremes because they could store their food, while increasing their vulnerability in other ways because of the reduced variety of sources from which food was taken. Nonetheless, the extent of such domination could not affect the weather.

Over the last few centuries, however, new technologies have appeared that facilitated the enticing vision of a society liberated from the contingencies and inconveniences of rainfall, heatwaves, storms and droughts. Air-conditioning, flood defence systems, improved techniques of irrigation (Varshney, 1995) and so on seemed to underpin the distinctly modern possibility of controlling the weather or at least its effects. These technologies constitute the infrastructures that Paul N. Edwards observes "reside in a naturalised background" (Edwards, 2003: 185). Many contemporary ways of life depends on them, "yet we notice them mainly when they fail, which they rarely do" (Edwards, 2003: 185). Horn suggests that the ability to determine temperature and humidity constitutes perhaps what is the oldest dream of humankind — the reverie of "a life without weather, without meteorological contingencies and surprises, extreme weather events, seasonal changes, or locally challenging climate conditions" (Horn, 2016: 234). As she goes onto explain, technology such as air conditioning has allowed the emergence of a universal standard of an "ideal climate", part of a picture of modernity in which humans are distinct from nature and in which society is wrenched away from climate (Horn, 2016: 237).

One crucial part of the infrastructure that Edwards refers to is weather forecasting. Taken for granted today, the first daily predictions were pioneered in 1861 by Robert Fitzroy, founder of what would become the Met Office in London. Fitzroy had been concerned primarily with issuing storm warnings for sailors, but then began to submit what he called "weather forecasts" for *The Times* (Moore, 2016). In Fitzroy's words: "Prophecies and predictions they are not — the term forecast is strictly applicable to such an opinion as is the result of scientific combination and calculation." (quoted in Moore (2016: 567)). The data gathered in order to measure and forecast the weather paved the way for the emergence of "climate" as an object that could be neutrally observed, carefully measured and tentatively predicted.

Yet before we move on to consider this scientific object in the next section, it is worth remembering that, as we hope to have briefly illustrated in this section, weather events can be read and responded to in very different ways. Societies have always been confronted with variable weather conditions and environmental disasters that may have both limited and augmented the possibilities for agricultural practices, cultural traditions and political institutions (Xiao *et al.*, 2015). But such social and cultural differences also offer divergent ways of interpreting their changing environments. A comparison of the late Ming dynasty

(1560–1644) and late Qing dynasty (1780–1911) in North China, for example, shows that the social response to similar periods of climate deterioration and disaster can vary greatly (Xiao *et al.,* 2015).

Taking seriously the different possibilities for the relation between society and climate bids us to ask: how might certain presuppositions limit our interpretation of weather events and environmental catastrophe? What lessons might be drawn from past responses to climatic emergencies (Behringer, 2010) and fluctuating conditions (Janik, 2013)? Can more holistic worldviews, such as Buddhism, offer a salutary alternative to understanding environmental crisis (de Silva, 1998)?

1.2 The Climate of Science

Following the introduction of various meteorological instruments in the seventeenth century, there was a move to use standardised and regulated observations to understand the weather. As we explore further in Chapter Two, "climate" emerged at the close of the nineteenth century as an object of science and climate research became an independent scientific discipline. This new branch of science provided tables, charts, maps and atlases of climatic required for planning purposes by traders, the military and colonialists. Meteorological measurements became standardised so that regional climates could be mapped, compiled and compared. "Order was imposed on seemingly chaotic weather" (Hulme, 2009: 6). In the scientific depiction, the tangible qualities of rain, wind and sun were overshadowed by the data on humidity, precipitation, temperature and so on. Climate and weather therefore should be understood as different: Weather is the transient, real, local and unruly atmospheric condition of a particular moment, and should be distinguished from the abstract entity of climate. In the scientific rendering, climate could only be accessed through statistics, usually calculated over long periods of time and large geographical areas. Climate, in other words, is the "statistics of the weather". Consider the description of climate given in the nineteenth century by Austrian meteorologist Julius von Hann (1839–1921):

"The totality of meteorological phenomena, which characterise the (average) condition of the atmosphere in any position of the earth's surface."

This is remarkably similar to the twentieth-first century description of climate provided by the US National Aeronautics and Space Administration (NASA):

"Climate is the description of the long-term pattern of weather in a particular area … When scientists talk about climate, they're looking at averages of precipitation, temperature, humidity, sunshine, wind velocity, phenomena such as fog, frost, and hail storms, and other measures of the weather that occur over a long period in a particular place."[1]

Here, climate is a matter of *average* conditions that do not exist in reality. Climate is a scientific construction, created through the collation of a series of measurements and observations of atmospheric values — primarily temperature, precipitation and wind speed.

The first meteorological measurement network of the *Societas Meteorologica Palatina* (1780–1792) was established in the 1780 by the Mannheim Academy of Science in Germany (Cassidy, 1985; Kington, 1964; Lüdecke, 1997). Using carefully selected locations and standardised instructions, it produced outstandingly reliable and extensive data. Its aim was to report monthly and yearly averages that might facilitate recognition of patterns in climate. Figure 1.1 shows, as an example, simultaneous air pressure readings at three locations in Europe, which display the eastward passage of a low-pressure system during December 1775.

Since then, methodological and technological developments have allowed knowledge of the climate system to develop and expand. Today satellite based observation systems have improved the monitoring of climatic processes. For example, the "Global Precipitation Measurement mission" (GPM) is a joint project of NASA

---

[1] https://www.nasa.gov/mission_pages/noaa-n/climate/climate_weather.html

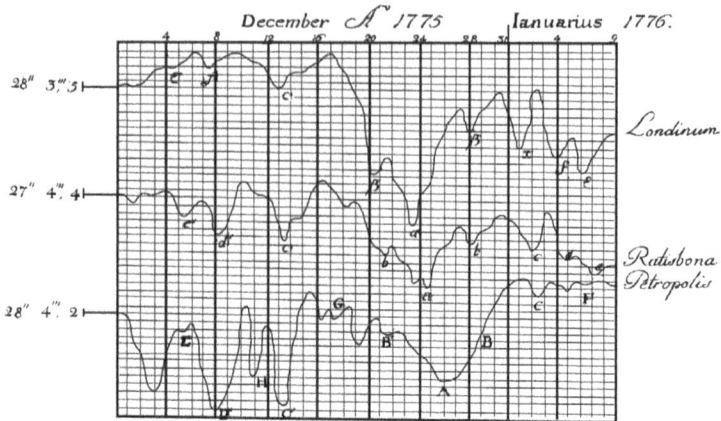

Fig. 1.1. Air pressure record for London, St. Petersburg and Regensburg, collected by the Societas Meteorologica Palatina for the period December 1775 to January 1776.

and the Japan Aerospace Exploration Agency.[2] GPM uses satellites to track precipitation worldwide every few hours and creates global precipitation maps in order to improve the knowledge of the earth's climate. This technology can help forecast tropical cyclones, floods and cyclones, freshwater availability in addition to daily weather patterns.

Sophisticated technology and advanced scientific research have facilitated more complex and sensitive analysis of the interaction of various components of climate. This has permitted the realisation of useful climate models, which can realistically describe natural processes and their sensitivity to various factors. These models simulate the earth's climate and act as experimental devices for climate research. Although numerous models exist, uncertainties remain and predictions are not watertight, not least because social behaviour and the release of greenhouse gas emissions is impossible to predict (Brown and Caldeira, 2017). For more on climate modelling see Chapter Two.

Alongside this technological change was a discernible paradigm shift; climate was no longer understood so much as *average weather*,

---

[2] https://www.nasa.gov/feature/goddard/how-nasa-sees-el-ni-o-effects-from-space

but as a global system that delineated *possible weather* (McGregor, 2006). In these scientific depictions of climate as a system working at a *global* level, the diversity of material, social and psychological impacts of climate that occurs at a *local* level is collapsed. Climate is essentially regarded as a *neutral object* that can be quantitatively described in terms of the measurement of climatological variables and their interaction (Hulme, 2009: 9).

As various commentators point out, and not only recently, the meanings of climate vary in ways that exceed this sort of description; global averages and global models cannot convey the full impact of the climate upon the intricacies of human cultures. Climate, in short, cannot simply be understood as a neutral scientific object but must be understood to be a matter of social and political concern — climate irresistibly affects society. It is in the shadows of this recognition that climate determinism finds its hold; climate determinism is the doctrine asserting that society is ultimately the effect of climatic dynamics.

As we discuss in Chapter Four, while climate determinists are right to detect the relevance of climate to society, they dangerously overlook the agency of human individuals and communities. Social agency becomes undeniable, however, with the widespread recognition of anthropogenic climate change. The dubious linear fatality of climate determinism is rumbled by the acknowledgement that climate, and climate science too, is implicated by the social contexts in which it operates. With this recognition, climate is no longer understood as entirely *natural* but is also partly *social*.

1.3 The Climate of Society

Climate variation is not a new topic of scientific research, and nor is its anthropogenic nature. As we will show in Chapter Three, during the eighteenth and nineteenth centuries, the variability of climate, and the implications of this for society, was acknowledged and discussed. But what is arguably new is the widespread acknowledgement of *permanent* change. Scientists reveal the potentially dramatic and irreversible hazards generated by a volatile and unpredictable climate system.

Furthermore, contemporary discussions of climate change forefront the concern that climate change is not entirely natural. Human beings, in their growing numbers and demand for resources have had an indisputable and enduring impact upon their environment through agricultural techniques, industrialisation, deforestation and exploration. Once known as "global warming", the phenomenon is not limited to rising temperatures but also involves melting glaciers, rising sea-levels and increased risks of hurricanes, typhoons, droughts, floods and wildfires. Emissions of greenhouse gases generated by human societies now overwhelm any "natural" external processes in affecting the climate (Hayhoe et al., 2017: 134).[3] It is this recognition that has drawn attention to the interconnection between society and climate.

The revelation of the extent of human involvement in nature has gone as far as the questioning of the very reality of a "nature" that is independent of society. "Nature", it is said, has come to an end (cf. McKibben (1990)). For what we normally understand to be "nature" is actually usually already altered by the activities and architectures of human society. Human beings have always lived amidst nature in remote parts of the planet. Human industry and agriculture have always borrowed from and plundered the land. More recent conservation practices actually designed to *preserve* features of the natural landscape, also serve to sculpt its contours. Our experience with the environment is rarely, if ever, that of a habitat in its pure natural state. Untouched nature is seldom encountered, for as Vincent Di Norcia writes:

"Nature would have already been mediated many times over through varied technologies of manufacture, transport, communication and settlement. Consequently, the immediate environment of members of such societies is technological, the product of multiple transformations of nature." (Di Nocia, 1974: 89–90)

Despite the persistent fascination held by images of a pristine "wilderness", untouched by human hand, this idea of the purity of nature is, at best, a naïve and romanticised myth. At worst, it is a denial

---

[3] we describe the Greenhouse Effect in more detail in Chapter Two

of both the human societies that have long co-habited such so-called "wilderness" and the growing impact of human society on delicately balanced ecosystems. Indeed, the extent of human impact is both pervasive and palpable. Biodiversity loss, pollution, an exponentially growing global population along with a changing climate and all of its associated dimensions and effects, are examples of the ways in which social systems are wreaking havoc upon natural systems, comparable to catastrophic events in the earth's history that have occurred only very, very rarely in the past (Kolbert 2014). The extent and depth of human impact on the geology of the planet has led some to proclaim the crossing of an "epoch-scale boundary" (Zalasiewicz *et al.*, 2011) and the dawning of a new geological epoch.

A group of scientists are arguing that human activity has left such an indelible mark upon the earth's ecosystems and atmosphere, that we have entered the "Anthropocene" (Purdy, 2015; Wapner, 2014). Introduced by atmospheric chemist Paul Crutzen and biologist Eugene Stoermer less than two decades ago (Crutzen and Stoermer, 2000), the central tenet of the Anthropocene is that, by examining the pollen record and the "reef gap" of sedimentary rock, geologists of the future will be able to detect a qualitative change in the stratigraphic record (Brondizio *et al.*, 2016). In short, the human species has turned itself into "a globally potent biogeophysical force, capable of leaving a durable imprint in the geological record" (Revkin, 2011).

Although its starting date and scientific status remain highly contentious (Monastersky, 2015; Certini and Scalenghe, 2015; Zalasiewicz *et al.*, 2015; Waters *et al.*, 2016), the Anthropocene has become a prominent topic of scientific research. From across the natural and social sciences and humanities, researchers are considering its impact upon the fundamental assumptions of their disciplines. The Anthropocene demands a renewed look at traditional notions of the subjects and objects of responsibility; conceptions of time (Dibley, 2012); understandings of history (Chakrabarty, 2009; Mauelshagen, 2014); modes of literature (Ghosh, 2016); the structures of governance (Lövbrand *et al.*, 2009); projects of environmentalism (Wapner, 2014); and constructions of knowledge (Machin, 2017). New academic disciplines are emerging in order to grasp the immense implications of an integrated "earth system" in which nature and society are intertwined. Earth system science is an "integrated system of knowledge" (Jackson, 2009: 596) which "aspires to understand the interactions and feedbacks among the components of the earth system, encompassing the lithosphere, atmosphere, hydrosphere, cryosphere, and biosphere as well as human societies and economies" (Jackson, 2009: 297) and on which "human society may depend in the face of global climate change" (Jackson, 2009: 596). Beyond the academy, the term is leaving its own mark, appearing in political commentary, media articles, environmental campaigns and museum exhibitions.[4]

Regardless of the debates over the validity of coining a new geological epoch, the insight of the Anthropocene for this book is that the climate can no longer be understood as entirely "natural" but is at least in part a result of social "interference". Climate is susceptible to human influence, manipulation and disruption; climate incorporates both natural and social influences, so that the boundary between the two becomes blurred, and it is impossible to understand the "natural" climate separately from social context. Human beings can hardly set themselves outside of the natural processes in which they are embedded; they are always both a *part of and apart from* their environments. The relationship between nature and society is not fixed; a multiplicity of possibilities exists for this relationship, depending upon both the environmental conditions and social imaginaries at work. Climate is both a component of this entanglement and its product.

This means that any attempt to forecast the future of climate and society is on unstable ground. The complexity of the relationship between the various morphing components of the earth system, comprised of geophysical processes along with social institutions and cultural practices, makes predictions difficult. This uncertainty ruptures

---

[4] "Welcome to the Anthropocene" is both the title of a special issue of *The Economist* in 2011 (see www.economist.com/node/18744401) and an exhibition at the Deutsches Museum in Munich, www.deutsches-museum.de/en/exhibitions/special-exhibitions/anthropocene/.

any easy presupposition about the future, so that prognoses vary wildly: hopes that nature will find its equilibrium sit uncomfortably alongside dystopian depictions of catastrophe, propelling very different types of social response to climate change and climate science (see Chapter Five).

The array of different possible responses to climate change has meant that a putatively "natural" condition has become a political concern. As we show in Chapter Six, climate has become a part of the political sphere; not just an object of scientific research but an issue of political debate at local, national and global levels, in various institutions and discourses of government and governance. The political climate has thus changed too. Not only is climate a new policy area, but existing policy areas — energy and transport, for example — have had to be adjusted.

Climate change has also meant a reaffirmation of the dependence of politics and politicians on expertise. We engage with the connection between policymaking and science in Chapter Five. To grasp the meanings and implications of climate change is to comprehend scientific data and methods, and to communicate scientific data to policy makers and the public is to be familiar with political processes. But this does not mean that climate science informs policymaking in a linear direction. Understanding climate change is not simply a matter of appreciating the scientific facts. Science has commonly been regarded to be separate from the processes it describes and equipped to generate objective knowledge and "speak truth to power". But this is a misleading picture (Turner, 2014: 4). Stephen Turner emphasises, "Scientists and experts have interests. Systems of expertise have biases ... expertise itself is dependent on other people's knowledge and on the systems that generate it." (Turner, 2014: 4)

The issue of climate change brings to attention the way in which science can never entirely free itself of the social context in which it is researched, funded and communicated. As Shelia Jasanoff points out, "Social processes colour the extent to which pieces of scientific knowledge are perceived as certain ... In areas of high uncertainty, political interest frequently shapes the presentation of scientific facts and hypotheses." (Jasanoff, 2012: 103)

As is far more acknowledged in eastern thought (de Silva, 1998), climate, society and science are tightly interconnected — society impacts the climate and depends upon science to understand and respond to this impact, while science is rooted in society that is itself conditioned by the climate. And unlike earlier times when weather extremes were interpreted as sorcery, today there are no witches to burn if things go wrong.

1.4 Conclusion: Climates of Disagreement

The ability of society to influence the climate has been taken by some to decry the destructive force of humanity and others to celebrate it. A celebration of human ingenuity coincides with the "promethean" construction of nature as something that can be *mastered* by a society unconstrained by environmental limits. In Greek myth, the Titan *Prometheus* stole fire from Zeus for the benefit of humanity and illustrated the ability of humans to manipulate the world. John Dryzek (2005: 52) uses the term "promethean" to denote "the unlimited confidence in the ability of humans and their technologies to overcome any problems — including environmental problems". Such confidence is perhaps most evident today in some of the expectations surrounding climate engineering projects, in which technologies of solar radiation management and carbon sequestration aim to reduce the amount of carbon dioxide in the atmosphere or to redirect heat from the sun back into space (Hulme, 2014; Ruser and Machin, 2016). This Promethean construction contrasts with the depiction of contemporary forms of human civilisation as fundamentally and perilously out of tune with their natural environment (cf. Kingworth and Hine (2009)) and the call for a drastic cutback in energy use and a dramatic change in society and economy (Klein, 2014). It contrasts with different worldviews in which society and nature are part of an interdependent matrix and that would therefore reject the possibility of a technological "quick fix" to environmental problems (de Silva, 1998).

As we have attempted to illustrate above, climate — and climate change — can be understood and treated by society in various ways. Climate can be a constraint, threat and resource for the human beings by whom it is utilised, feared, protected and imagined. Different conceptions of climate exist within and across different cultures and ways of life and correspond to different conceptions and experiences of "nature".

This means, in short, that the scientific research, social meaning and political governance of climate is no straightforward matter. Climate is something about which we will inevitably disagree (cf. Hulme (2009) and Machin (2013)). In the rest of this book we consider the various themes, meanings, questions and disagreements of climate in more detail. We discuss climate as a scientific object and a background condition of everyday life, a source of catastrophe, a social determinant and as a policy concern. We argue that while climate impacts our social reality in multiple and intersecting ways, different social conceptions of climate jostle for position.

References

Behringer, Wolfgang (1999) "Climatic change and witch-hunting: The impact of the Little Ice Age on mentalities," *Climatic Change* **43**: 335–351.

Behringer, Wolfgang (2010) *A Cultural History of Climate*. Oxford, Malden: Polity Press.

Brondizio, Eduardo S., Karen O'Brien, Xuemei Bai, Frank Briemann, Will Steffen, Frans Berkhout, Christophe Cudennec, Marian C. Lemos, Alexander Wolfe, Jose Parma and Olivira Chen-Tung (2016), "Re-conceptualising the Anthropocene," *Global Environmental Change* **39**: 318–327.

Brown, Parick and Ken Caldeira (2017) "Greater future global warming inferred from Earth's recent energy budget," *Nature* **552**: 45–50.

Campbell, Bruce (2016) *The Great Transition: Climate, Disease and Society in the Late-Medieval World*. Cambridge: Cambridge University Press.

Cassidy, David (1985) "Meteorology in Mannheim: The Palatine Meteorological Society, 1780–1795," *Sudhoffs Archive*, 8–25.

Certini, Giacomo and Riccardo Scalenghe (2015) "Is the Anthropocene really worthy of a formal geologic definition," *The Anthropocene Review* **2**(1): 77–80.

Chakrabarty, Dipesh (2009) "The climate of history: Four theses," *Eurozine*.

Crutzen Paul J. and Eugene F. Stoermer (2000) "The Anthropocene," *The Global Change Newsletter* (41): 17–18.

de Silva, Padmasiri (1998) *Environmental Philosophy and Ethics in Buddhism*. Hampshire and London: Macmillan Press.

Di Norcia, Vincent (1974) "From critical theory to critical ecology," *Telos* **22**: 85–95.

Dibley, Ben (2010) "'The Shape of Things to Come': Seven theses on the Anthropocene and attachment," *Australian Humanities Review* 52.

Dryzek, John (2005), *The Politics of the Earth: Environmental Discourses*. Oxford: Oxford University Press.

Edwards, Paul N. (2003) "Infrastructure and modernity: Force, time, and social organisation in the history of sociotechnical systems". In: *Modernity and Technology*. (Eds.) Thomas J. Misa, Philip Brey and Andrew Feenberg, Cambridge, MA: MIT Press, pp. 185–225.

Frazer, James (2002) [1890, 1st ed.] *The Golden Bough: A Study in Religion and Magic*. New York: Dover Publications.

Ghosh, Amitav (2016) *The Great Derangement: Climate Change and the Unthinkable*. London and Chicago: University of Chicago Press.

Hayden, Brian (2009) "The proof is in the pudding: Feasting and the origins of domestication," *Current Anthropology* **50**(5): 597–601.

Hayhoe, K., J. Edmonds, R. E. Kopp, A. N. LeGrande, B. M. Sanderson, M. F. Wehner and D. J. Wuebbles (2017) "Climate models, scenarios, and projections". In: *Climate Science Special Report: Fourth National Climate Assessment, Volume I*. (Eds.) Wuebbles, D. J., D. W. Fahey, K. A. Hibbard, D. J. Dokken, B. C. Stewart and T. K. Maycock. US Global Change Research Program, Washington, DC, USA, pp. 133–160.

Horn, Eva (2016) "Air conditioning: Taming the climate as a dream of civilisation". In: *Climates: Architecture and the Planetary Imaginary*. (Eds.) Graham, James, Caitlin Blanchfield, Alissa Anderson, Jordan Carver and Jacob Moore. New York: Columbia Books on Architecture and the City, pp. 233–241.

Hulme, Mike (2009) *Why We Disagree About Climate Change: Understanding Controversy, Inaction and Opportunity*. Cambridge: Cambridge University Press.

Hulme, Mike (2014) *Can Science Fix Climate Change? A Case Against Climate Engineering*. Cambridge: Polity

Humboldt, Alexander von (1856) [1845, 1st ed.] *Cosmos, A Physical Description of the Universe*. (Trans.) E. C. Otte. New York: Harper and Brothers.

Humboldt, Alexander von (1849) [1808, 1st ed.] *Aspects of Nature: In Different Lands and Different Climates*. (Trans.) Mrs. Sabine. Philadelphia: Lea and Blanchard.

Janik, Liliana (2013) "Changing paradigms: Flux and stability in past environments," *The Cambridge Journal of Anthropology* **31**: 85–104.

Jackson, Stephen T. (2009) "Alexander von Humboldt and the General Physics of the Earth," *Science* **324**: 596–597.

Jasanoff, Sheila (2012) *Science and Public Reason*. London: Routledge.

Jordan, William Chester (1996) *The Great Famine: Northern Europe in the Early Fourteenth Century*. Princeton University Press.

Kington, John A. (1964) "The Societas Meteorologica Palatina: An eighteenth-century meteorological society," *Weather* (29): 416–426.

Kingsnorth, Paul and Dougald Hine (2009) *Dark Mountain Manifesto*. Available at: http://dark-mountain.net/about/manifesto/

Klein, Naomi (2014) *This Changes Everything: Capitalism vs. the Climate*. New York: Simon and Schuster.

Kolbert, Elizabeth (2014) *The Sixth Extinction: An Unnatural History*. London, New Delhi New York and Sydney: Bloomsbury.

Kvideland Reimund and Henning. K. Sehmsdorf (eds.) (1991) *Scandinavian Folk Belief and Legend*. University of Minnesota Press, Minneapolis.

Latour, Bruno (2017) *Facing Gaia: Eight Lectures on the New Climatic Regime*. Cambridge, Medford: Polity Press.

Leiss, William (1972) *The Domination of Nature*. New York: George Braziller.

Lewis, Mark Edward (2006) *The Flood Myths of Early China*. New York: State University of New York Press.

Lövbrand, Eva, Johannes Stripple and Bo Wiman (2009) "Earth system governmentality: Reflections on science in the Anthropocene," *Global Environmental Change* (19): 7–13.

Lucas, Henry (1930) "The Great European Famine of 1315, 1316, and 1317," *Speculum* **5**(4): 343–377.

Lüdecke, Cornelia (1997) "The monastery of Andechs as station in early meteorological observational networks," *Meteorologische Zeitschrift* (6): 242–248.

Machin, Amanda (2013), *Negotiating Climate Change: Radical Democracy and the Illusion of Consensus*. London: Zed Books.

Machin, Amanda (2017) "Sustaining democracy: Science, politics and disagreement in the Anthropocenem". In: *Nachhaltigkeitswissenschaften und die Suche nach neuen Wissensregimen*. (Ed.) Thomas Pfister. Munich: Metropolis.

Mauelshagen, Franz (2014) "Redefining historical climatology in the Anthropocene," *The Anthropocene Review* **1**(2):171–204.

McKibben, Bill (1989) *The End of Nature*. New York: Random House.

Meinert, Carmen and Claus Leggewie (2013) "Foreword". In: *Nature Environment and Culture in East Asia*. (Ed.) Carmen Meinert. Leiden and Boston: Brill.

Monastersky, Richard (2015) "Anthropocene: The human age," *Nature* (519): 144–147.

Moore, Peter (2016) *The Weather Experiment: The Pioneers who Sought to see the Future*. London: Vintage.

Neiman, Susan (2015) *Evil in Modern Thought: An Alternative History of Philosophy*. Oxfordshire, New Jersey: Princeton University Press.

Ogilvie, Astrid E. J. and Gísli Pálsson (2003) "Mood, magic, and metaphor: Allusions to weather and climate in the sagas of Icelanders". In: *Weather, Climate, Culture*. (Eds.) Strauss, Sarah and Ben Orlove. Oxford, NY: Berg.

Oster, Emily (2004) "Witchcraft, weather and economic growth in renaissance Europe," *Journal of Economic Perspectives* **18**(1): 215–228.

Purdy, Jedediah (2015) *After Nature: A Politics for the Anthropocene*. Cambridge Massachusetts: Harvard University Press.

Revkin, Andrew (2011) "Confronting the 'Anthropocene'," *New York Times*. 11 May. Available at: http://dotearth.blogs.nytimes.com/2011/05/11/confronting-the-anthropocene/ (accessed: 14 March 2018).

Ruser, Alexander and Amanda Machin (2016) "Technology can save us, can't it? The emergence of the 'techno-fix' narrative in climate politics". In: *Technology + Society =? Future. MASA Conference Proceedings*.

Stehr, Nico (1978) "Man and the environment: A general perspective," *ARSP*, LXIV/1.

Stehr, Nico and Hans von Storch (2000) "Global warming in perspective. Contemporary concern about climate change has an age-old resonance," *Nature* **405**: 615.

Turner, Stephen P. (2014) *The Politics of Expertise*. Routledge.

Varshney, R. S. (1995) "Modern methods of irrigation," *GeoJournal* **35**(1): 59–63.

Wapner, Paul (2014) "The changing nature of Nature: Environmental politics in the Anthropocene," *Global Environmental Politics* **14**(4): 36–54.

Waters, Colin N., *et al.* (2016) "The Anthropocene is functionally and stratigraphically distinct from the Holocene," *Science* **351**(6269).

Wu, Qinglong, *et al.* (2016) "Outburst flood at 1920 BCE supports historicity of China's Great Flood and the Xia dynasty," *Science* **353**(6299): 579–582.

Xiao, Lingbo, *et al.* (2015). "Famine, migration and war: Comparison of climate change impacts and social responses in North China between the late Ming and late Qing dynasties," *The Holocene* **25**(6): 900–910.

Zalasiewicz, Jan, *et al.* (2015) "When did the Anthropocene begin? A mid-twentieth century boundary level is stratigraphically optimal," *Quaternary International* **383**: 196–203.

Zalasiewicz, Jan, Mark Williams, Alan Haywood and Michael Ellis (2011) "The Anthropocene: A new epoch of geological time?" *Philosophical Transactions of the Royal Society A*. (369): 835–841.

CHAPTER 2
Climate as Scientific Object

Everybody talks about the weather, but nobody does anything about it.

Charles Dudley Warner, 1897[1]

The weather, its vicissitudes and its extremes are a constant source of concern and fascination. For the weather impacts directly upon everyone; it influences our behaviour, our mood, our health and our plans. Climate, however, is importantly distinguished from the weather: "Unlike the wind which we feel on our face or a raindrop that wets our hair" Mike Hulme writes "climate is a constructed idea that takes these sensory encounters and constructs them into something more abstract" (2009: 3). As the National Aeronautics and Space Administration (NASA) website puts it: "Climate is what you expect… and weather is what you get".[2]

This means that although the palpable *effects* of climate have a very real and direct impact, climate can only be *indirectly* experienced. Climate is somehow derived from the totality of weather occurrences. How then does this derivation work? The everyday construction of climate is different from the scientific construction; we might call them "typical weather" and "average weather". From a lay perspective, climate might be perceived as "typical weather" for a certain location. Typical weather might include very dry summers in some years, but

---

[1] Charles Dudley Warner was a journalist and editor at the *Hartford Courant* toward the end of the nineteenth century; the quoted phrase appeared in an editorial of the *Hartford Courant* in August 1897. But it is a contested matter whether or not the phrase originated with Mark Twain, who was a friend and neighbour of Warner's.

[2] https://www.nasa.gov/mission_pages/noaa-n/climate/climate_weather.html

very wet ones in others. "Typical weather" is a useful construction, but not necessarily scientifically robust.

Climate scientists, in contrast, have utilised the notion of "average weather", which is not the same thing. As Julius von Hann (1839–1921) explained in his *Handbook of Climatology*, considered by his contemporaries to represent the state of the art: "Climate science will ... have the task of acquainting us with the average atmospheric conditions over the different parts of the earth's surface." (von Hann, 1903: 2). "Average weather" is a constructed mathematical artifact that does not exist in reality. Unlike "typical weather" it cannot allow for the occurrence of erratic weather. In both depictions, however, climate translates the locally experienced weather into an overarching construction.

In this chapter we start by considering some of the statistics generated by climate science and explaining how they are useful for society. We examine the scientific artefact of the climate as "average weather". This artefact emerged at the end of the nineteenth century when scientific methods and instruments began to produce a reliable numerical representation of observable climate variables. We then show a further shift away from the analysis of individual climate variables towards a global analytical approach in which climate is conceptualised as a complex and variable interactive system in which the diverse components of ocean, atmosphere and ice work with and against each other. Here climate is not *average* weather, nor *typical* weather, but rather *possible* weather.

2.1 Climate as Average Weather

Just as climate is not the same thing as weather, climatology is not the same thing as meteorology. Meteorology deals primarily with the physics of atmospheric processes and is commonly associated with weather forecasting. Weather forecasting usually has a maximum time-horizon of 10 to 15 days (Wynne, 2010: 294). Predictive skill has been steadily improving due to advances in technological acumen, but forecasts nonetheless are given on a scale of hours, days or weeks (Bauer *et al.*, 2015). Climatology, in contrast to meteorology, is concerned with much longer time periods. Initially, the task of climatology was to determine the statistics of weather (consisting of average conditions and frequency of extreme events, and so on) in the various different regions of the world. Thus, we might say that climatology was considered meteorology's bookkeeper (McGregor, 2006: 1). The two disciplines are therefore intimately interlinked since climate predictions draw upon meteorological data (Bauer *et al.*, 2015).

It is climatology then that provides the details of the two cycles that frame our everyday life: these cycles work regionally on different time scales: diurnal (daily) and annual (yearly). The diurnal cycle involves, for example, temperature and humidity changes over the day. Figure 2.1 shows as an example of the "diurnal" cycles for summer conditions in Germany. Maximum temperatures are reached at about 14:00 and minimum temperatures not before 6:00. The temperature contrast in the coastal resort of Warnemünde is only about 5°C, and thus much smaller than at inland Potsdam (near Berlin), with a daily temperature range of about 7°C.

In the annual cycle, the rise and fall of air temperature from month to month leads to the differentiation of the seasons. The monthly mean average annual cycles of rainfall and the monthly mean average day and night temperatures for a number of stations throughout the world, are given in Figures 2.2 and 2.3. These figures illustrate the striking differences in the climates in different parts of the globe.

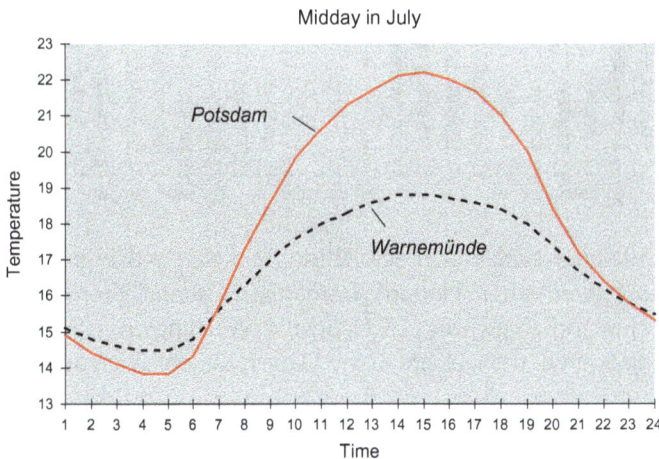

Fig. 2.1. Mean diurnal cycle of temperature at two stations in Germany in July: Warnemünde at the Baltic Sea coast and at Potsdam, close to Berlin. Source: Friedrich-Wilhlem Gerstengarbe and Peter Werner.

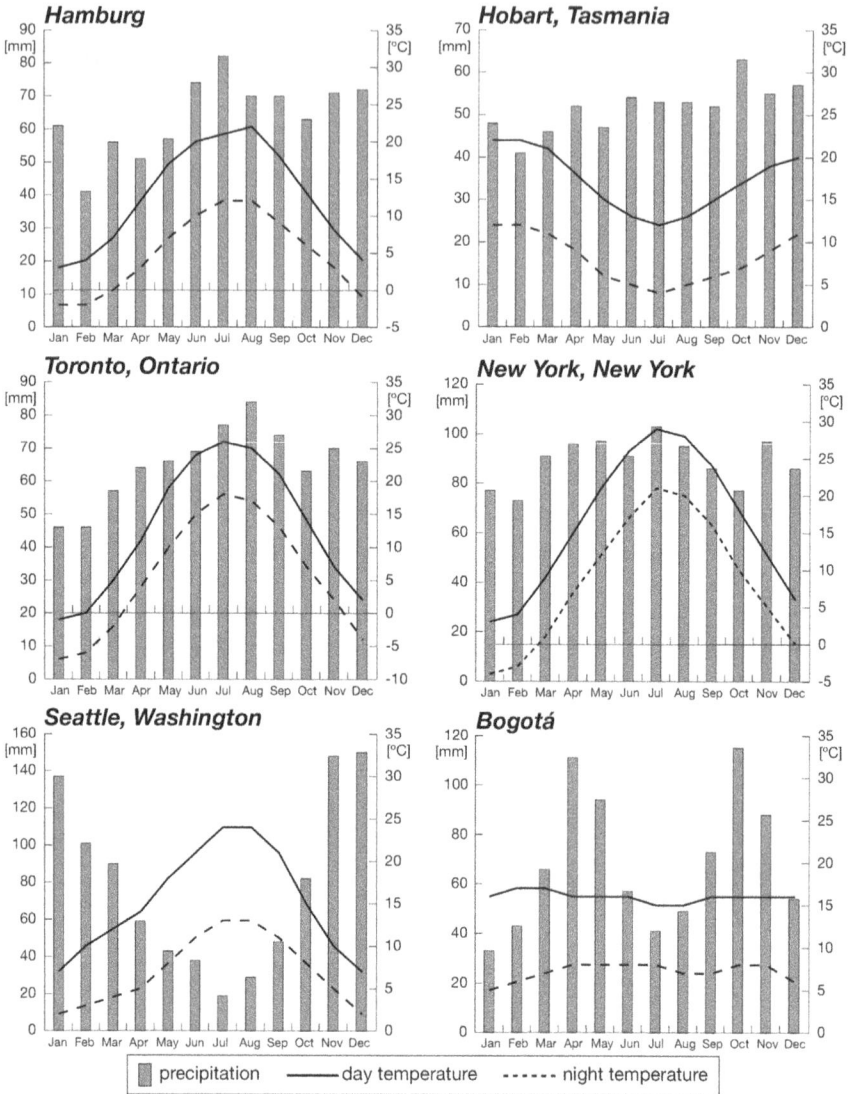

Fig. 2.2. Annual cycles of precipitation, and day and night temperature in Hamburg (Germany), Hobart (Tasmania, Australia), Toronto (Ontario, Canada), New York (New York, USA), Seattle (Washington, USA) and Bogotá (Colombia). Source: the website of the Danish Meteorological Institute.

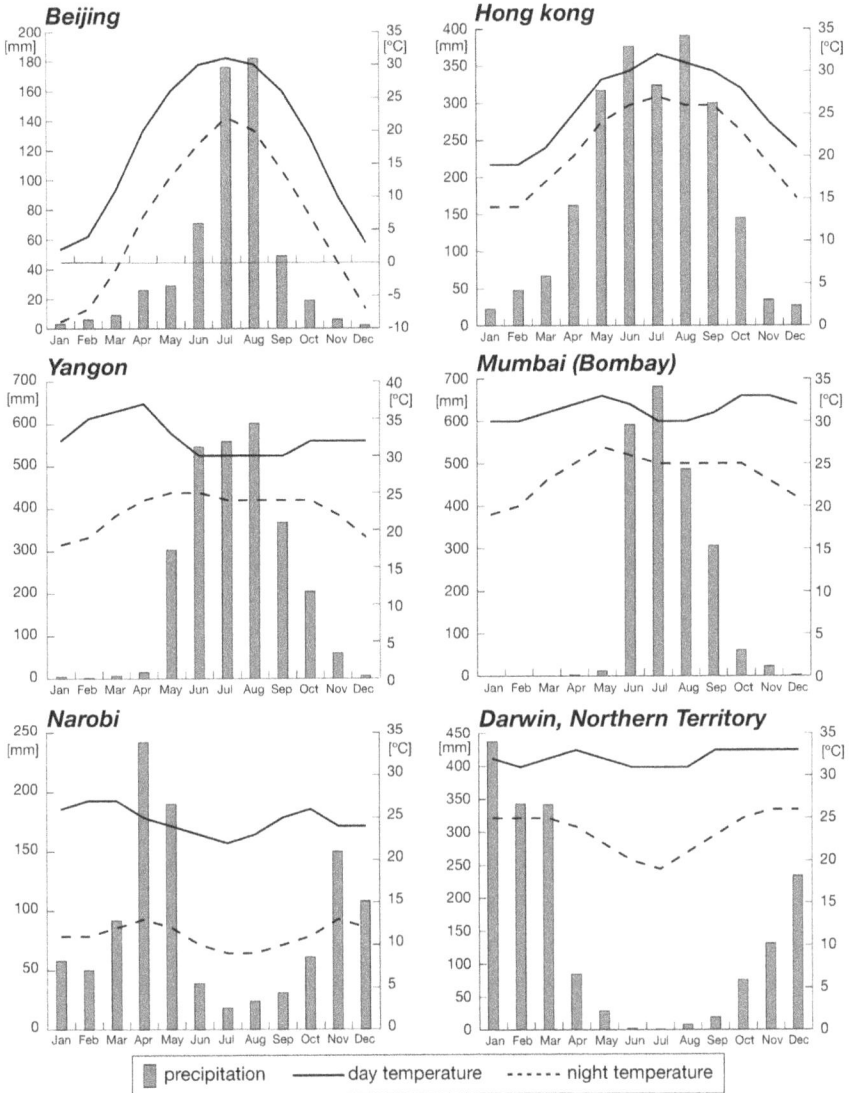

Fig. 2.3. Annual cycles of precipitation, and day and night temperature in Beijing (China), Hong Kong (China), Yangon (Myanmar), Mumbai (Bombay, India), Nairobi (Kenya), Darwin (Northern Territory, Australia). Source: the website of the Danish Meteorological Institute.

The coldest and warmest months are generally understood to be distinguished as winter and summer. Actually, it is important to note that the official seasons are determined *astronomically*, namely as the three-month period between the shortest day and the next equinox (when day and night are of equal length). This determination differs by about three weeks from the *meteorological* definition.

The division of the seasons, as far as it is based on the observations of the temperature range, is not identical with the length of the days. In the earth's temperate climate zones, the greatest cold is not recorded on the shortest day of the year, nor the greatest warmth on the longest. This is due to "seasonal lag" which arises because it takes some time for the earth's surface to heat up. This means that surface temperatures will be rising even though (solar) energy from the sun is decreasing, and conversely, that temperatures will fall even when solar energy is increasing.

Also shown in Figures 2.2 and 2.3 is the absence of a pronounced annual temperature cycle in the tropical regions that surround the equator. In the temperate climate zones of the northern and southern hemispheres, we can separate four annual seasons more or less clearly from one another. Mumbai, Yangdon and Darwin, in contrast, show a semi-annual oscillation, with two maxima and two minima each year. Figures 2.2 and 2.3 also show the mean annual course of precipitation amounts, which varies greatly among the locations shown. Some show a notable dry period and a marked wet period, or monsoon season (such as Mumbai or Yangon); others show an almost uniform precipitation amount (such as Hamburg, Hobart or New York). A "bimodal" distribution, with two maxima and minima, can also be found among the cases, namely Nairobi and Bogotá.

Besides surface air temperature and precipitation as the most important and widely used climatic variables, meteorological services regularly record other variables, such as humidity, wind, air pressure, and the number of hours of sunshine. Another climatic variable that is taken not from meteorological but rather from hydrographical services is the water level along coasts, lakes and rivers.

This sort of data may seem somewhat unremarkable today. But it should not be taken for granted. Not least because of its crucial connection to the planning and preparation of policy decisions, particularly in the context of potential climatological risks, hazards and benefits. Policymakers and planners commonly request and refer to the information generated by climatologists and meteorologists to make important decisions that will affect almost everyone's lives and livelihoods.

For example, statistics from climatology, meteorology and hydrology are combined with those of hydraulic engineering and geographic data on topography, infrastructure and population to estimates flood risks in particular regions vulnerable to flooding (Morss *et al.*, 2005: 1594). Flooding can take human lives and cause huge damage to property. In coastal areas, growing population and property development alongside potential climate changes exacerbate the threat of flood hazards (Morss *et al.*, 2005: 1593). Statistics play a role in the decision-making regarding flood defence strategies.

Statistics on indicators such as storm intensities and the frequency of cyclones are also invaluable. Tropical cyclones (also called hurricanes or tropical storms) are one of the most violent weather phenomena on the planet (Ramsay, 2017). Climatologists have found a remarkably steady rate of 80 tropical cyclones per year, which occur disproportionately around the planet (70 per cent of all tropical cyclones occur in the northern hemisphere). Since the 1960s when meteorological satellites became available, scientific knowledge has grown enormously on these phenomena. Figure 2.4 shows the frequency of the formation of cyclones in different regions of the world.

Scientific data on cyclones is commonly presented in the forms of maps (see Figure 2.5). Such information can be used, along with socio-economic data and local knowledge, to inform policymaking for defensive strategies for vulnerable populations. Improved defensive measures include cyclone shelters, evacuation plans, coastal embankments, reforestation schemes, awareness programmes and educational campaigns (Haque *et al.*, 2012).

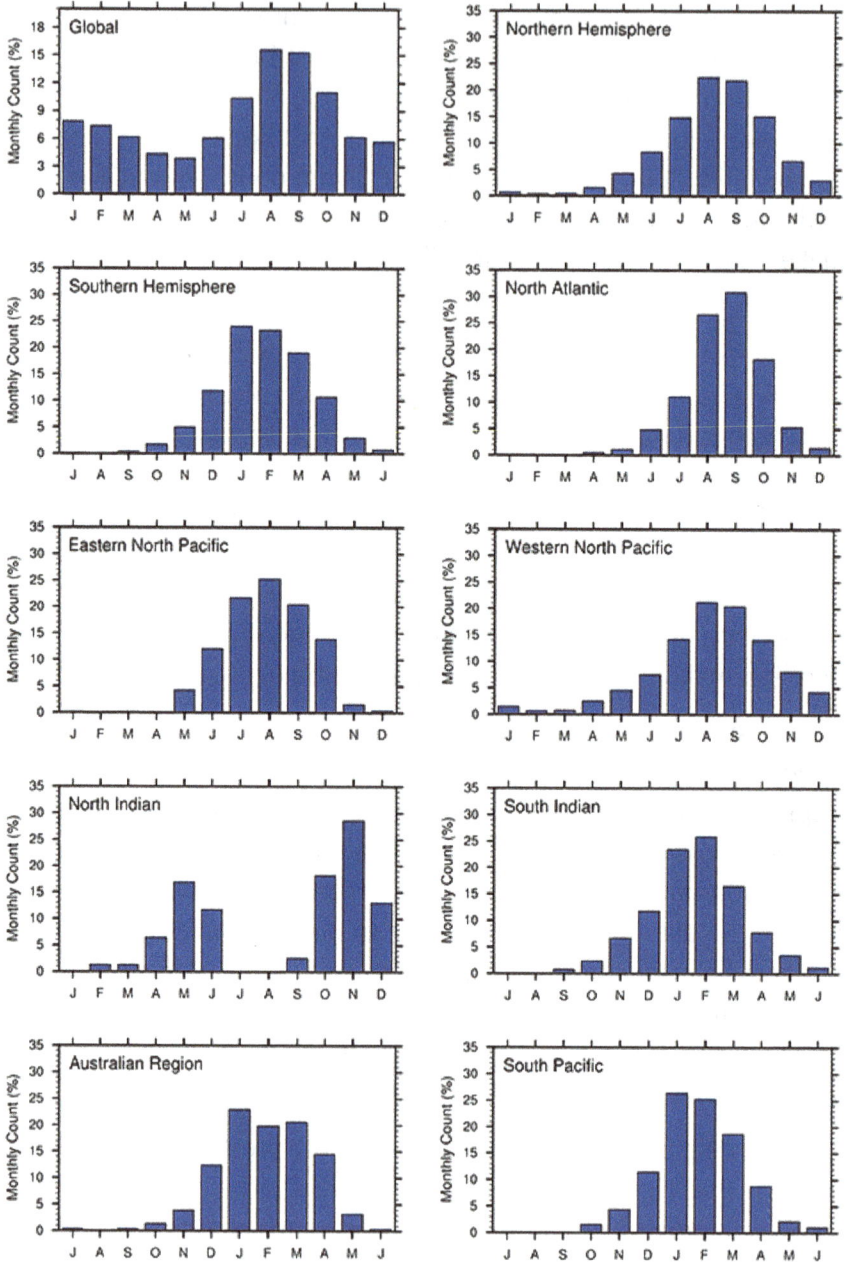

Fig. 2.4. Monthly frequency of tropical cyclone formation count for the globe, each hemisphere and seven individual basins, based on data from 1985 to 2014. Note that the Southern Hemisphere annual cycle begins in July and ends in June. Source: Ramsay (2017).

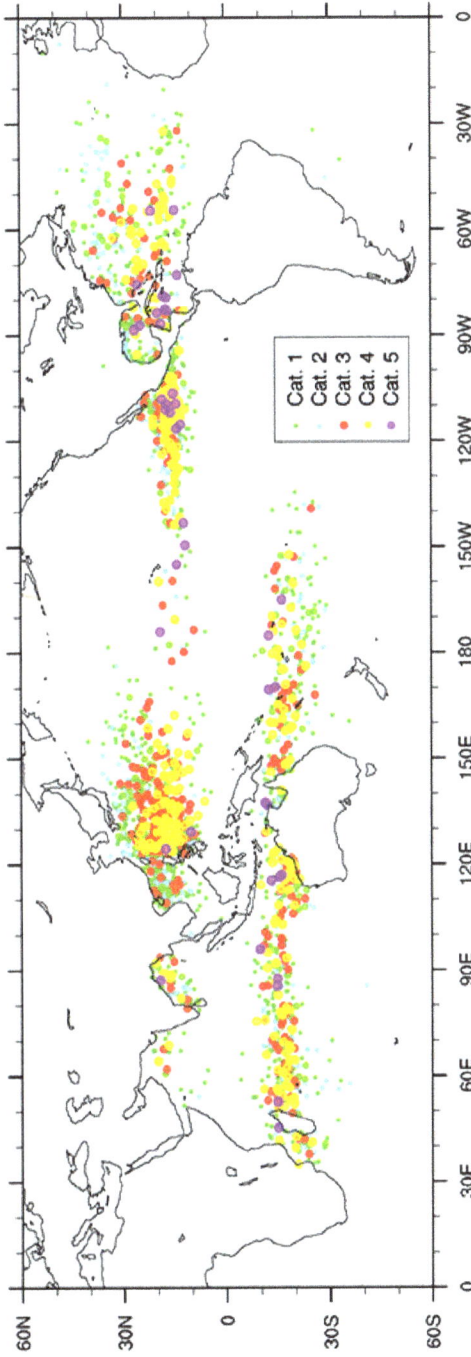

Fig. 2.5. The locations of lifetime maximum intensities (LMI) of tropical cyclones for the period 1985–2014. LMI is color-coded according to category on the Saffir–Simpson Hurricane Wind Scale. Source: Ramsay (2017).

Climate statistics are also invaluable for the agricultural industry, since agriculture is particularly vulnerable to weather extremes (Shannon and Motha, 2015). Different types of agriculture will be suitable for different particular regions with different temperatures, humidity, precipitation and risk of weather extremes. For some crops, it is not only the average summer or winter temperatures that are important, but also the most extreme cold or the timing of the beginning of the frost period.

In Florida, for example, frost wipes out an entire citrus crop, which is extremely sensitive to weather patterns (Shannon and Motha, 2015: 53); climate is the most important component in determining the quality of citrus production (Zekri, 2011: 6). Relatively wide annual and diurnal temperature differentials are key to producing good fruit (Miller and Glantz, 1988). Florida's semitropical climate (alongside economic and social factors) has been a crucial reason for the emergence and success of the huge citrus industry, but it appears that climate — and, more importantly, the perception of climate risks — may be playing a role in its relative decline, and the concomitant rapid development of Brazil's orange juice industry (Miller and Glantz, 1988). Agricultural practices, such as the use of irrigation and fertilizers and risk management strategies, can mitigate some of the effects of climate, but not all (Zekri, 2011: 8). In recent years, however, Florida's orange harvest has actually been harder hit by extreme weather events such as hurricanes, as well as disease. Scientists use climate data to make forecasts on crop yields that are important for making decisions that affect orange farmers and traders.[3] As Shannon and Motha observe, input from the farmers themselves is crucial for preparing effective strategies (Shannon and Motha, 2015: 50).

However, the way in which scientific statistics are used in political decision-making is hardly straightforward. The idea that scientific knowledge is simply presented to policymakers in a linear process

---

[3] See CEC, http://water.columbia.edu/research-themes/risk-and-financial-instruments/oranges-and-climate-predicting-the-usda-citrus-forecast-error/

is misleading; rather science interacts in a context characterised by uncertainty, disagreement and the competition between different priorities and perspectives. As Morss and colleagues found in their study of the role of scientific statistics in flood policy, the production and communication of more detailed scientific information does not necessarily make decision making easier or quicker. "Practitioners in floodrisk estimation and management make decisions in complex settings characterised by significant change and uncertainty. They also operate under regulatory, institutional, political, resource and other constraints that limit their capacity to use new scientific information." (Morss *et al.*, 2005: 1594). Moreover, the idea that climate statistics are easily calculated and that they can offer a complete and neutral picture of reality is deeply flawed, as we consider next.

2.2 Measuring the Climate

An important issue that concerns scientists and those who rely on their data, is the matter of the methods by which the sorts of statistics considered in the previous section are generated. Producing valid statistics depends upon the collation of observation data provided by various instruments and reliable methods over a period of time. The standardisation of meteorological measurements over the eighteenth and nineteenth centuries led to the replacement of undetermined climatic observations with more precise numerical specifications. This meant that, as demonstrated above, instead of *descriptive* statements, climate can be described *numerically* in terms of average rainfall and maximum temperatures. The scientific study of climate led to a reliable representation of observable climate variables and thus to a numerical language.

As we refer to in Chapter One, the *Societas Meteorologica Palatina* was established as a permanently funded international network of meteorological observers in 1780. It was the precursor of an era in which "new empirically-based predictions would replace the long tradition of unsubstantiated farm rules and folk sayings" (Cassidy, 1985: 9).

Alongside the inestimable benefits of this scientific numerical approach to climate arose significant challenges. One fundamental concern was methodology: which of the many quantitative measurements available should be used and how could they be standardised? There was a practical need for a limited number of robust measurements that were unaffected by non-meteorological processes, that were relevant for their intended application, and were representative for an area and a segment of time of interest.

Climate researchers sought methods to measure the relevant climate variables so that the numerical values could be reproduced for the place in question, and in addition also permit a comparison with other locations. This task was not easy. As a simple illustration take, for example, the quantity "daily average temperature", which is derived by averaging several observations throughout the day. The result depends on the time the thermometer reading is taken. Switching the reading times from 6:00, 12:00, 18:00 and 24:00, to 7:00, 13:00, 19:00 and 1:00 will produce different "daily average temperatures".

Not only is it not always clear which type of data is relevant and representative and practical, but the data itself is subject to possible inaccuracies due to measurement inconsistencies and ambiguities. The history of meteorology and oceanography knows many "inhomogeneities" of observation data, generated by changing observational practices.

Consider the measurement of the sea surface temperature, a crucial source of data for climate scientists. These temperatures were first taken, as early as 1853, by measuring the temperature of seawater hauled to the deck in a bucket. Later however, a different method was used; measurements were taken from the seawater pumped through the engines of steamships to cool them. From the 1920s, ship engineers began observing intake seawater temperature in the cooling water for monitoring purposes. Meteorologists recognised that these intake temperatures might be useful (Matthews, 2013). However, procedures and instruments were hardly standardised and remain poorly documented today (Matthews, 2013: 685); intake depths vary and ships themselves have changed. Moreover, there appear to be

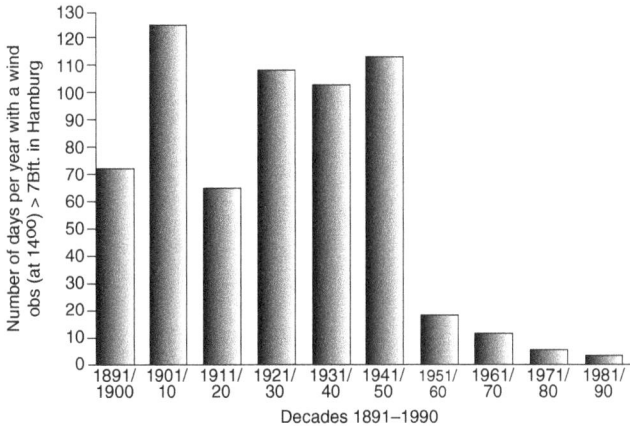

Fig. 2.6. Frequency of recorded gale wind conditions in Hamburg per decade. Source: Heiner Schmidt.

discrepancies between the two methods of data collection (Matthews, 2013: 691).

Or consider the apparent fall in occurrences of strong wind in Hamburg. Figure 2.6 shows the number of days with wind strength 7 Beaufort and more within 10-year periods. The resulting numbers showed more than 90 of these occurrences in the decades before 1951–1960; since then, only 10 more occurrences were noted. But the reason for this variation has nothing to do with any climatic changes, but rather with the transfer of observation sites from the maritime weather office in the harbour to the airport. The observations are known to be correct, but they are evidently not representative of Hamburg's storm climate.

Or consider the data on tornadoes in the US (see Figure 2.7). Before 1870, reports about the occurrence of tornadoes were sporadic and anecdotal. Only in the 1870s did the Signal Service of the US Army begin to systematically collect reports. However, this activity was viewed as politically unfavourable, since there was concern these violent events would scare away potential immigrants (cf. Gutmann and Field (2010)). Therefore, tornado activity was significantly underreported in the late 1880s, and this was only rectified a few years later.

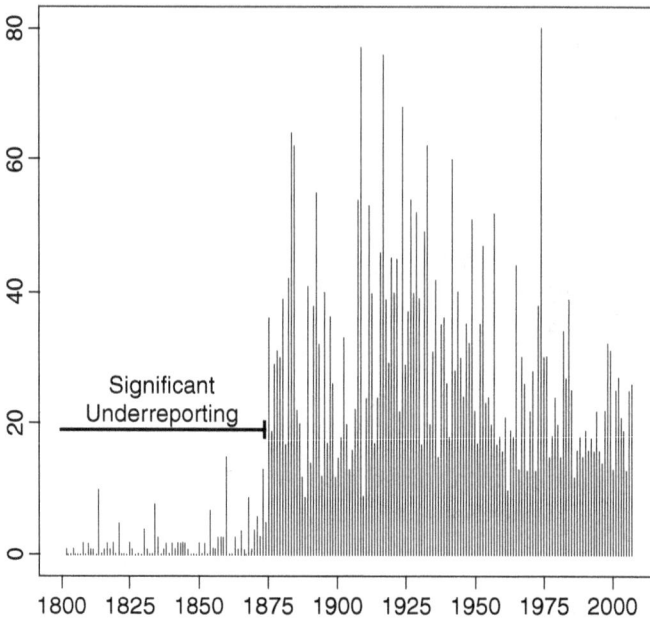

Fig. 2.7.   Frequency of tornado reports in the US. From Harold Brooks.

Just as it is important to notice the potential mismeasurements of climate, it is also important to notice some of the limitations of the conventions of climate science. For example, because there are no "natural" definitions for the periods of time for which data should be averaged, meteorological services agreed on a standard observational period of 30 years; averages for many climate statistics are therefore based on a temporal aggregation over 30 years. However, not only has it become clear that climate also shows notable variations for different time scales (see Chapter Four), but such temporal aggregation result in the overlooking of dynamic processes such as species establishment (Serra-Diaz *et al.*, 2016).

Our point here is not simply that one should be wary of claims of scientific objectivity and accuracy. It is also that statistics — and the interpretations of statistics — utilise and reaffirm various contingent assumptions that might be brought to attention. Problematic assertions of the certainty of numerical data can be unpicked by critics to undermine scientific claims (Machin and Ruser, 2018). This is not to say, of course, that climate statistics do not play a valuable, perhaps even an invaluable, role for society. As we consider next, however, the notion of "climate as average weather" is now overshadowed by a new paradigm.

2.3 The Emergence of a New Paradigm

In the nineteenth century the scientific global object of climate was understood as consisting of the accumulation of *regional* descriptions (von Storch, 1999). Then, most scientists interested in climate were geographers who had regarded climate as a regional issue, and were focused upon its limiting and conditioning effect for the regional fauna, flora and human society. It was only later that scientists began to look at the system *as a whole*. As Phillip Lehmann describes, scientists began to demand "new deductive approaches and more narrowly defined data sets, increasingly drawn not from *around* but from *above* the earth and focusing on atmospheric dynamics rather than historical and geographical information" (Lehmann, 2015: 58). As early as 1895, Clevland Abbe demanded that climate scientists turn away from an eclectic mix of historical and geographical data towards a focus on atmospheric dynamics. This comprised a significant shift: "climate scientists… started to move away from the messy concerns of the earth and geography and into the atmosphere" (Lehmann, 2015: 69).

The great thirst for greater knowledge of the giddy heights above the earth's surface both took advantage of, and fuelled, technological advances. In the nineteenth century, observations of the atmosphere had been somewhat tethered to the ground. Scientists undertook difficult expeditions with unwieldy equipment to take measurements from the tops of mountains and spectacular manned balloon flights. In 1787, for example, Horace Benedict de Saussure climbed the summit of Mont Blanc carrying a thermometer, barometer, telescope, compass and other instruments, explaining in his diary that he "was bound to make the observations and experiments which alone gave value to my venture" (Freshfield, 1920: 232).[4] Saussure was

---

[4] In 1784 Benedict de Saussure also measured the depths of Lake Constance.

not the first man to reach the summit, but from the measurements he gathered, it was understood that the temperature of the earth's atmosphere dropped by about 0.7°C per 100 metres of altitude. It was assumed that temperature would always decrease with height, approaching absolute zero.

By the early twentieth century, however, it became possible to obtain observation data from much higher altitudes, with the help of unmanned weather balloons and kites. The French meteorologist Leon Teisserenc de Bort was one of these "pioneers of the upper air" (Shaw, 1913: 519) and a member of "the school that regards the meteorology of the globe in its entirety as a condition for effective progress" (Shaw, 1913: 520). Using *balloon sondes* Teisserence de Bort was able to take measurements from heights of up to nearly 30 kilometres. These measurements were puzzling as they revealed that temperature stayed the same beyond a certain altitude. A layer of the atmosphere in which there was no change in temperature had been discovered, which Teisserenc de Bort named the "stratosphere" in a paper written in 1908. The German scientist Richard Assmann is credited with making this discovery at the same time (Nebeker, 1995: 48). It is in the stratosphere, 35 kilometres thick, that the ozone layer is found; ozone absorbs solar radiation, which causes the temperatures to be higher than might be expected (Charlson, 2006: 137).

The stratosphere was one factor in a total complex and interactive system that comprises the climate. In the new — and still prevailing — climate paradigm, the climate is thus understood as a *global system,* itself "an integral part of the biogeophysical system earth" (Nicholson, 2017: 4). The number and relative significance of climate variables changed in this paradigm. In the past, individual climatic features were analysed in an isolated manner. Today, in contrast, scientists try to include numerous diverse variables to describe an integrated climate system. As McGregor explains, in climate system theory: "Climate is the manifestation of the interaction among the major climate system components of the atmosphere, hydrosphere, cryosphere, biosphere and land surface, and external forcings, such as solar variability and long-term earth–sun geometry relationships." (McGregor, 2006: 1).

Below we offer a brief description of some of these components before turning to the Greenhouse Effect.

The atmosphere is the layer of gas that surrounds the earth. Its main constituents are nitrogen, oxygen, argon, water and carbon dioxide. Short-wave radiation from the sun heats the atmosphere, especially in the tropics. The air near the earth's surface becomes significantly warmed and the air layer columns become unstable; low-lying air becomes lighter than the air above. In this way, vertical air transport movements are formed. These are amplified by the fact that when air ascends, it expands, cools off and can therefore retain less water vapour, so that a part of the gaseous water becomes fluid again. During this condensation, the energy originally needed to evaporate the fluid water into water vapour is released as heat. This additional energy warms the rising air, which thereby becomes lighter again than its surroundings and can therefore continue its ascent. From an airplane in the tropics, one can observe this process in the powerful cloud towers that often exceed 11,000 or more metres, often above the flight altitude.

The hydrosphere consists of the parts of the earth composed of water. In addition to the oceans, which comprises 71 per cent of the earth's surface, water is found in rivers, groundwater, glaciers and ice sheets, soil, biomass, and as we noted above, the atmosphere. The hydrosphere, long appreciated as essential to life, plays a central role in the climate system. Water is the primary transporter of heat as it flows through the hydrologic cycle, driven by solar radiation. The ocean functions as a primary heat reservoir, regulating global temperatures. Oceans also absorb roughly half of the carbon dioxide emitted into the atmosphere (Henshaw *et al.*, 2006: 114).

The cryosphere is part of the hydrosphere that is frozen, consisting of ice caps, glaciers and sea ice. The cryosphere contains the largest reserves of freshwater on Earth, although the exact quantity is difficult to estimate (Henshaw *et al.*, 2006: 114). It has been a fairly static component, but it plays two important roles in the dynamic climate system. First, it isolates the ocean and the soil from the atmosphere, so that the exchange of heat and moisture will be reduced. Also it contributes to albedo, meaning that it reflects short-wave solar radiation back from the earth's surface.[5]

The biosphere is unique (it is believed!) to Earth. It is made up of the parts of the earth where life exists, the "biota"; from the bottom of the ocean to the highest mountain peaks and it includes human beings too. The biosphere is not geographically distinguishable, but rather exists *within* the other spheres (Jacobsen *et al.*, 2006: 4). Previously conceived to be merely responsive to climate the biosphere is now regarded as an integral part of the system (Jacobsen *et al.*, 2006: 6).

The climate system, then, is a result of the interactions between these distinct but interconnected systems characterised by energy flows, feedbacks and non-linearity (McGregor, 2006: 1). This global climate system is hardly static. The interplay of the different components of the climate system generates variations at all time scales and at different geographic scales. Small causes may quickly initiate large effects, as in a "butterfly effect"; the flap of a butterfly's wing can radically change the development of the entire system. But in the climate system there is not just one butterfly, but rather millions of butterflies beating their wings uninterruptedly. This poses serious challenges for climate scientists.

2.4 The Discovery of the "Greenhouse Effect"

An important piece in this picture of a global system is the "Greenhouse Effect". Jean-Baptiste Joseph Fourier (1768–1830) is commonly, although erroneously, cited as having first discovered this mechanism; he certainly never used the term "Greenhouse" (or "Serre" in French) (Flemming, 1999). Fourier might be credited with having established the problem of planetary temperatures as an object of scientific study (Pierrehumbert, 2004), but it was three quarters of a century later that Swedish chemist Svante Arrhenius (1859–1927) established the physics of the process of the "Greenhouse Effect". Arrhenius, who would later

---

[5] Albedo is the property of a surface to reflect short-wave radiation, measured in per cent. Deserts and snow have a high albedo; forests and seas have a low albedo.

receive the Nobel Prize in chemistry for other achievements, started with the idea that the incoming short-wave (solar) radiation would be balanced by the outgoing long-wave (thermal) radiation emitted from the earth. If it did not balance, the temperature would fall or rise until this balance was reached. Were a vacuum to exist between the sun and the earth, the average air temperature of the earth would be around −10°C. But this is clearly not the case. The reason for this is the earth's atmosphere.

The earth's atmosphere contains gases. Some of these gases, including water vapour, carbon dioxide and methane are "greenhouse gases" that catch long-wave radiation originating from the surface of the earth (and from the atmosphere itself) and radiate it back in all directions. Thus, part of the energy radiated from the ground does not escape directly to space — as would be the case if there were no atmosphere — but is partially returned to the ground and lower levels of the atmosphere after having been absorbed and re-emitted by the greenhouse gases. Greenhouse gases work in this way at very low concentrations. Water vapour is the most abundant and efficient greenhouse gas, although its atmospheric concentration varies from 0.2 per cent to 4 per cent. Carbon dioxide makes up only 0.03 per cent of the atmosphere.

Only about 40 per cent of heat radiation "gets through" to space, while 60 per cent of the energy is radiated back. Not only the short-wave solar radiation, but also the long-wave heat radiation returned from the atmosphere arrives at the ground. If we assume that the earth system normally has the above-mentioned temperature of −10°C, then it would indeed warm up as it gathers energy. The warming causes an increase in intensity of the long-wave radiation, of which, again because of the greenhouse gases, only 40 per cent continuously reaches space. Because the radiation grows with the temperature, the temperature increase causes more long-wave energy to be emitted. The warming exhausts itself when the 40 per cent of the radiated long-wave energy that reaches space compensates for the solar radiation arriving at the earth's surface. The "final temperature" that results is considerably higher than the original −10°C. However, because the atmosphere not only absorbs and re-emits long wave radiation, it also shields the ground to some extent against incoming solar radiation. Only a portion of the solar radiation found at the top of the atmosphere reaches the ground. The shielding depends on the albedo. The combined effect is moderated so that an average temperature of about 15°C finally appears, which indeed corresponds to observation. This, then, is the "Greenhouse Theory", actually a confusing designation because the temperature in a glass greenhouse is warmer than that of the surrounding air for other reasons.[6]

What is striking about Arrhenius's theory is the fact that today, 100 years after its publication, it is still acknowledged to be correct in an almost unchanged form (Arrhenius, 1896: 237–276). Arrhenius calculated the rise in air temperature in the case of a doubling of the atmospheric carbon dioxide concentration, and found a value comparable with present estimates of about 3°C. He maintained that a doubling was certainly possible, but only in 1,000 or more years, because 85 per cent of the carbon dioxide emitted into the atmosphere would accumulate in the ocean.[7] Today we know that the absorption in the ocean takes place over a protracted period of time and therefore a doubling of the CO_2 concentration in several decades is quite possible, and even very probable.[8]

---

[6] This description is somewhat simplified. A number of other processes, such as convection, modify the picture.

[7] See the remarkable textbook: S. A. Arrhenius, *Das Werden der Welten* (Leipzig Akademische Verlagsanstalt m.b.H., 1908). Or, in English, *Worlds in the Making: The Evolution of the Universe*. In this book Arrhenius described many aspects of the climate system correctly and comprehensibly, although he failed to describe the functioning of the sun properly, as he had no knowledge of nuclear processes driving the sun. Instead, he speculated about obscure chemical processes.

[8] Other notable efforts that aim to explain the general atmospheric circulation were initiated by the Englishman George Hadley (1685–1768) in the seventeenth century. Even though only very little empirical information was available, and in particular no data beyond the boundary layer of the atmosphere, he grasped essential parts of the general circulation correctly, such as the trade wind system. But he could not deduce other significant parts at that time. The German philosopher Immanuel Kant (1724–1804) also worked in this field of study; he analysed wind observations from ships in Southeast Asia, drawing from them the conclusion that there must be a continent further in the South, which was at that time still unknown Australia.

In the early twentieth century, other climate researchers made important contributions to physically-orientated climate research, such as the Norwegian Vilhem Bjerknes, who made some major breakthroughs, which included a new model of the extratropical cyclone (Friedman, 1989), the Swede Carl Gustav Rossby, who devised a mathematical theory of turbulence (Nebeker, 1995: 80), and the American John von Neumann, who after the Second World War recognised the possibilities of electronic data calculations for weather predictions and implemented the first applications of the newly developed computers (Halmos, 1973).[9]

2.5 Modelling the Climate

Another important and related change in the shift to a new climate science paradigm was the change in research agenda. The primary purpose of climate studies is no longer to *collect and analyse* many detailed observations in order to facilitate various planning objectives. Instead, it is to *model and predict* the climate system; the role of observations has been reduced to the validation of models and to the refutation of hypotheses. Investigations into climate are no longer *descriptive*, but are rather primarily *analytical*.

An important tool used by scientists are climate models which have become "the central scientific currency of the field" (Wynne, 2010: 293). Climate models, under construction since the 1960s, attempt to incorporate as many as possible of the relevant components and interactions of the climate system. Using supercomputers these models serve as a virtual manipulatable "quasi-reality" that allows for the conducting of experiments that could not take place in reality. They are thus neither "true" nor "false"; but rather are "adequate" or "inadequate" and "useful" or "not useful" (Müller and von Storch, 2010: vii).

---

[9] Von Neumann was not just interested in predicting the weather but controlling it. He proposed research into the idea of dyeing the polar icecaps so as to decrease the amount of energy they would reflect (Halmos, 1973: 393)

Models are always, of course, *approximations* of the actual climate system. As Peter Müller and Hans von Storch explain, climate models are "quasi-realistic" (Müller and von Storch, 2010: 20). In contrast, the climate system in reality is, of course, an "open" environmental system, meaning that it is exposed to a series of uncontrollable, external influences that are impossible to specify and account for in a model. External factors include astronomical and social influences and others that may not even be known. Thus, while certain aspects of the climate system are satisfactorily (but not perfectly) represented in climate models, other are not. In particular, thermodynamic climate processes (such as the formation of clouds) present challenges. These processes are generally well understood if one operates on small spatial scales. But in climate models, the smallest resolved spatial scales are many orders of magnitude larger than the microphysical scales of these thermodynamic processes. An example: the process of absorbing and reflecting radiation in the atmosphere is decisive for the formation of the climate. In this regard, clouds are of particular importance: the process of absorption and reflection of radiation depends on the size of the water droplets in the clouds and the effect can be significant. For the climate model, however, the size of the drops is irrelevant. Such processes will thus be only approximated or "parametrised". This means that models can never *describe reality* but only portray parts of it. Nonetheless, contemporary "quasi-realistic" models are generally considered to be able to describe the climate system well enough to be major tools in climate science (Müller and von Storch, 2010: 6). We consider climate models further in relation to their role in understanding climate change in Chapter 4.

Another concern, however, is that climate models, depending on the degree of complexity, tend to only show structures that describe features with an extent of over at least many hundred, or even a thousand, kilometres. From the perspective of these models, regional and local variations are subsumed under the analysis of climate at a *global* level. Only the very largest structures are of importance for the shaping of the global climate. For example, the disappearance of Australasia in its entirety would of course change the climate of that continent but would not significantly affect global climate. From this perspective, the regional climate is the global climate modified by regional details such as types of land use, regional mountain ranges, marginal oceans and large lakes. Furthermore, local climates originate out of regional climates through adaptation to local details, such as large cities, small lakes (e.g., Lake Constance) and small mountain ranges (e.g., the Appalachian Mountains).

As a rule, regional and local climates are inadequately simulated in climate modelling. So although these models are successful in simulating the *global climate* they may be unable to offer much analysis on the regional or local level (Urry, 2011: 28). And yet it is precisely at a regional and local level that human populations experience the climate.

2.6 Conclusion: Climate as Possible Weather

Ultimately, the ability of scientists to predict *the weather* is limited. The weather is the reason why we are often interested in climate; climate gives us an idea of what weather to expect, what crops to plant, the houses to build, the clothes to pack; it provides a framework for arranging our lives, a general picture to refer to when we make plans for the future.

The scientific object of climate as "average weather" provides statistics, charts and maps that are ultimately an abstraction, but offer a fascinating picture of the different components and manifestations of the environment of different parts of the planet. As we have described, however, a paradigm shift led to the emergence of the scientific object of climate as a "global system", constituted by various tightly and complexly interconnected sub-systems. Climate is no longer *average weather*, instead climate has become something that *delimits possible weather*. As McGregor writes, in this paradigm, climate is not "the statistical assemblage of the weather at a location or region". Instead "climate describes the conditions under which 'things' are possible" (McGregor, 2006: 1).

What complicates matters for scientists in their attempts to understand the global system of climate, is that the different components of this system each undergo variation and interconnect in different ways with other morphing components. The climate system is characterised by change, replete with unpredictability. As we go on to discuss in Chapter Three, climate change is not a new topic of scientific research; climate has long been understood as something that varies, and we outline the different types of variation and the challenges that this poses for both science and society.

References

Arrhenius, Svante A. (1908) *Das Werden der Welten*. Leipzig: Akademische Verlagsgesellschaft.

Arrhenius, Svante A. (1896) "On the influence of carbonic acid in the air upon the temperature of the ground," *Philosophical Magazine and Journal of Science* **41**: 237–276.

Bauer, Peter, Alan Thorpe and Gilbert Brunet (2015) "The quiet revolution of numerical weather prediction," *Nature* **525**: 47–55.

Charlson, Robert J. (2006) "The atmosphere". In: *Earth System Science: From Biogeochemical Cycles to Global Change*. (Eds.) Jacobsen, Michael C., Robert J. Charlson, Henning Rodhe and Gordon H. Orians. Elsevier.

Flemming, James (1999) "Joseph Fourier, the 'greenhouse effect', and the quest for a universal theory of terrestrial temperatures," *Endeavour* **23**(2): 72–75.

Freshfield, Douglas W. (1920) *The Life of Horace Benedict de Saussure*. London: Edward Arnold.

Friedman, Robert M. (1989) *Appropriating the Weather. Vilhelm Bjerknes and the Construction of a Modern Meteorology*. New York: Cornell University Press.

Gutmann, Myron P. and Vincenzo Field (2010) "Katrina in historical context: Environment and migration in the US," *Population and Environment* **31**: 3–19.

Halmos, P. R. (1973) "The Legend of John Von Neumann," *The American Mathematical Monthly* **80**(4): 382–394.

Haque, Ubydul, *et al.* (2012) "Reduced death rates from cyclones in Bangladesh: What more needs to be done?" *Bulletin of the World Health Organisation* **90**(2): 150–156.

Henshaw, Patricia C., Robert J. Charlson and Stephen J. Burges (2006) "Water and the Hydrophere". In: *Earth System Science: From Biogeochemical*

Cycles to Global Change. (Eds.) Jacobsen, Michael C., Robert J. Charlson, Henning Rodhe and Gordon H. Orians. Elsevier.

Jacobsen, Michael C., Robert J. Charlson and Henning Rodhe (2006) "Introduction: Biogeochemical cycles as fundamental constructs for studying earth system science and global change". In: *Earth System Science: From Biogeochemical Cycles to Global Change*. (Eds.) Jacobsen, Michael C., Robert J. Charlson, Henning Rodhe and Gordon H. Orians. Elsevier.

Lehmann, Philipp N. (2015) "Whither climatology? Brückner's climate oscillations, data debates, and dynamic climatology," *History of Meteorology* **7**: 49–70.

Machin, Amanda and Alexander Ruser (2019) "What counts in the politics of climate change? Science, scepticism and emblematic numbers". In: *Science, Numbers and Politics*. (Ed.) Markus Prutsch. Palgrave MacMillan.

Matthews, J. B. R. (2013) "Comparing historical and modern methods of sea surface temperature measurement," *Ocean Science* (9): 683–694.

McGregor, Glenn R. (2006) "Climatology: Its scientific nature and scope," *International Journal of Climatology* **26**: 1–5.

Miller, Kathleen A. and Michael H. Glantz (1988) "Climate and economic competitiveness: Florida freezes and the global citrus processing industry," *Climatic Change* **12**(1988): 135–164.

Morss, Rebecca E., Olga V. Wilheli, Mary W. Downton and Eve Gruntfest (2005) "Flood risk, uncertainty, and scientific information for decision making: Lessons from an interdisciplinary projectm," *American Meteorological Society*, 1593–1601.

Müller, Peter and Hans von Storch (2010) *Computer Modelling in Atmospheric and Oceanic Sciences: Building Knowledge*. Berlin, Heidelberg and New York: Springer.

Nebeker, Frederik (1995) *Calculating the Weather: Meteorology in the 20th Century*. San Diego: Academic Press.

Nicholson, Sharon (2017) "Evolving paradigms of climatic processes and atmospheric circulation affecting Africa". In: *Oxford Research Encyclopedia of Climate Science*. Oxford University Press.

Pierrehumbert, Raymond (2004) "Warming the world Greenhouse effect: Fourier's concept of planetary energy balance is still relevant today," *Nature* **432**.

Ramsay, Hamis (2017) "The global climatology of tropical cyclones". In: *The Oxford Research Encyclopedia of Natural Hazard Science,* doi: 10.1093/acrefore/9780199389407.013.79.

Serra-Diaz, Joseph M., *et al.* (2016) "Averaged 30-year climate change projections mask opportunities for species establishment," *Ecography* **39**(9): 844–845.

Shannon, Harlan D. and Raymond P. Motha (2015) "Managing weather and climate risks to agriculture in North America, South America, Central America and the Caribbean," *Weather and Climate Extremes* (10): 50–56.

Shaw, W.N. (1913) "Leon Philippe Teisserenc De Bort" *Nature* **90** (2254): 519–520.

Urry, John (2011) *Climate Change and Society*. Cambridge and Malde: Policy Press.

von Hann, Julius (1903) [1883, 1st ed.] *Handbook of Climatology. Volume 1: General Climatology*. New York: Macmillan.

von Storch, Hans (1999) "The global and regional climate systems". In: *Anthropogenic Climate Change*. (Eds.) von Storch, Hans and Götz Flöser. Berlin, Heidelberg: Springer.

Wynne, Brian (2010) "Strange weather, again: Climate science as political art," *Theory Culture & Society* **27**(2–3): 289–305.

Zekri, Mongi (2011) "Factors affecting citrus production and quality," *Citrus Industry* 2011: 6–9.

CHAPTER 3
Climate as Cycle and Change

It is generally remarked by people who have resided long in Pennsylvania and the neighbouring Colonies, that within the last forty or fifty years there has been a very observable Change of Climate, that our winters are not so intensely cold, nor our summers so disagreeably warm as they have been... The salutary effects which have resulted from cleaning and paving the streets of Philadelphia, are obvious to every inhabitant.

Hugh Williamson, 1770: 272, 280

We're stepping out onto a minefield and we don't know exactly where those mines are, but the farther we step out into the minefield — the more we warm the planet — the more likely it is that we do set off these mines, that we do encounter devastating tipping point-like changes in the climate.

Michael Mann, cited in Wernick, 2017

The dangers posed by a volatile climatic system, replete with tipping points, has perhaps at no time been as widely acknowledged as it is today, when the possibilities of anthropogenic climate change are the topic of numerous academic research projects, political campaigns and everyday discussions. And yet, the analysis of a long series of observations shows that significant climatic variations have always occurred. The presence of climate variability going back hundreds of years is neatly illustrated, for example, by the choice of the name "Greenland" by the Vikings in the Medieval Warm Period. From the eleventh to the thirteenth centuries, Greenland was still green due to the mild climate. Today it seems unlikely that anyone would choose such a name for this chilly land, covered to a large extent, currently at least, by permanent ice sheets. Data shows that temperatures dropped by more than a degree from the onset of the Little Ice Age in the thirteenth century (Klintisch, 2016). Scholars have suggested that one of the reasons (along with other unrelated economic issues and conflicts) for the disappearance of the Norse settlements in this region was due to the changes in the climate that disrupted the settlers' agricultural and hunting practices and trade (Dugmore *et al.*, 2007).

But the story doesn't end there. Ironically, just as scientists are uncovering the story and impact of climate change in Greenland, climate change is once again threatening Norse settlements. Organic artefacts preserved for centuries in the frozen soil — clothes and bones — are rapidly decaying as rising temperatures melt the permafrost. As one historian remarks: "It's horrifying. Just at the time we can do something with all this data, it is disappearing under our feet." (Poul Holm, quoted in Klintisch (2016)). Furthermore, as the permafrost thaws, the decaying organic matter releases the greenhouses gases carbon dioxide and methane into the atmosphere, thus contributing to a feedback loop of global warming (Adamson *et al.*, 2017).

This example, which we will return to later, raises some of the issues we will touch on in this chapter. If climate change occurred in the past, then in what ways is the contemporary situation unique? What are the different processes that provoke changes in the climate and how can they be distinguished? What are the implications for society? How have scientists understood climate variations in the past and how might their work be relevant in understanding the challenges faced by scientists and society today? And is it possible for climate change to be measured and predicted?

Our starting point is the historical discussion of climate change. As is apparently often forgotten, scientific discussions of climate variation date back (at least) to the nineteenth century. Here we consider the seminal, if overlooked, work of Eduard Brückner. For Brückner, however, climate variation was generally understood to take place cyclically. Today, in contrast, climate change is understood to be "progressive": irreversible and exponential. In the second section we turn to the contemporary scientific account of climate change, looking both at "natural" climate variability and then the mechanisms of anthropogenic (human caused) climate change. Finally, we pose the question of the uneven impact of both climate change and its diagnosis, across and within societies. This notion of climate as a variable, unpredictable and differentiated constraint on human social existence juxtaposes uneasily aside the notion of climate as a reliable stable object that we considered in Chapter Two. Our main point here is that to truly grasp its unique character, challenges and impact, anthropogenic climate change should be placed within its social and historical context, just as Brückner recommended.

3.1 The History of the Science of Climate Change

The intensive and often controversial discussion about anthropogenic global climate change today is not limited to scientific laboratories and esoteric journals, but engages politicians, the media and the general public. The "Greenhouse Effect" may arguably even be described today as having become common knowledge. But while it might appear that the topic of the human impact upon the climate has only just emerged, this is actually not the case. The interest in climate changes and their causes and consequences in geological or historic time is not unprecedented in science or in politics.

The issue of a changing climate was discussed, for example, in relation to the water levels of the Caspian Sea, the world's largest inland water body, which have shown substantial fluctuation over the past several hundred years (Chen *et. al.*, 2017). Systematic data collection began in 1830, but, reportedly, levels have noticeably varied for centuries. Scientists have debated for a long time whether the altered water level was a result of human activities, or brought about by a natural climate variation (Mekhtiev and Gul, 1996: 80).

Large-scale deforestation and the cultivation of large stretches of land were suspected to be human activities that affected the climate. The American physician, historian, statistician and "father of American forestry" Benjamin Franklin Hough, for example, wrote a report entitled: "The influence of forests upon climate" in 1875.[1] Hough was

---

[1] For a bibliography of Franklin B. Hough see the New York State Library entry, available at http://www.nysl.nysed.gov/msscfa/sc7009.htm

concerned about the status and conditions of forests, and the effect of their destruction upon climate. At the 1873 meeting of the American Association for the Advancement of Science (AAAS) Hough presented a paper entitled *On the Duty of Governments in the Preservation of Forests*, in which he argued that an excessive harvesting of trees in the Mediterranean had harmed the environment, and that a similar problem faced North America. As a result of Hough's paper, the AAAS formed a committee to educate Congress and state legislatures on the dangers of deforestation, and to recommend legislation (Steen, 1991).

However, human impact on climate has not always been seen as detrimental. One of the oldest documented scientific discussions on the topic of climate change considered the beneficial climatic effects of the cultivation of territory in North America during the eighteenth century to expound the purportedly salutary influence of colonisation. Hugh Williamson, a physician and scholar born in Philadelphia, reported in 1770 on climatic changes in Pennsylvania and its neighbouring regions of New England, stating that "the general improvement of the colonies have already produced very desirable effects" (Williamson, 1770: 280). He believed that the climate had actually improved due to the settlement of the region through colonisation, concluding that:

> "... clearing and smoothing the face of a country, would promote the heat of the atmosphere, and in many cases would prevent or mitigate those winter blasts which are the general origin of cold, whence the winters must become more temperate, and as facts appear to support and confirm our reasoning on this subject, we may rationally con-clude that in a series of years, when the virtuous industry of posterity shall have cultivated the interior part of this country, we shall seldom be visited by frosts or snows, but may enjoy such a temperature in the midst of winter, as shall hardly destroy the most tender plants." (Williamson, 1770: 277)

Williamson believed that "cultivation" of the land had led to a lessening of the terrible northwest storms and frosts in winter and this had had positive effects on agriculture and human health. This is certainly not a lone example of the way in which colonists "made knowledge about weather and climate and put it to use in the administration and inhabitation of colonial spaces" (Mahony and Endfield, 2018: 2).

The idea that climate was a physical entity that could *change* over time marked a paradigm shift (Mauelshagen, 2018). And once this idea of climate change had been born, it could be projected into the past to understand, for example, the changes in Greenland that had rendered it uninhabitable by ice. Although this idea emerged in the eighteenth century, it only became fully practical once data sets could be provided, which did not happen until the middle of the nineteenth century. Scientists in the nineteenth century found it increasingly evident that climate is not constant, but rather undergoes significant changes within centuries and decades.

One of the prominent protagonists of this climate change discussion over a century ago was German scientist Eduard Brückner, a professor of geography at the University of Vienna (see Figure 3.1).[2] "The climate fluctuates … and with it fluctuate rivers, lakes and glaciers," he wrote in 1889. (Brückner, 1889: 75). Brückner's main contribution to the discussion was his monograph, *Climate Changes since 1700*, published in 1890. The ambitious ideas laid out in this volume were crucial to, and indicative of, the emergence of a critical period in climatology that has informed the study of climate today (Berger *et al.*, 2002; Lehmann, 2015; Stehr and von Storch, 2000) and may therefore throw some new light on the contemporary discussion. We can perhaps learn from the holistic approach to climate change pursued by Brückner, who drew from various disciplines to understand the climate. Not only was he an early proponent of climate research and policy, he was also well aware of the impact of a changing climate, which was for this scientist, "not just a physical phenomenon to be described by scientists, but a

---

[2] Many of the articles by Eduard Brückner have been translated into English and published as an anthology edited by Nico Stehr and Hans von Storch, 2000: *Eduard Brückner — The Sources and Consequences of Climate Change and Climate Variability in Historical Times*. Kluwer Academic Publisher.

Fig. 3.1. Eduard Brückner, together with his teacher and colleague Albrecht Penck, was instrumental in detecting traces of previous ice ages in the European Alps. Source: Stehr and von Storch (2000).

powerful force with deep socio-economic and cultural repercussions" (Lehmann, 2015: 58). He felt ethically obliged to disseminate his research results, demanding political action on the basis of scientific evidence (Edwards, 2013: 67). In lectures and newspaper articles he addressed the public as well as professional groups such as farmers, who would be especially affected by climate changes. His ideas were discussed in the contemporary press. Thus, there is much in common between nineteenth-century and present-day climate change discussions. The important difference is that the earlier research was primarily concerned with *periodicity*.

In his work, Brückner presented the idea of climate change as occurring as part of a *cycle*. He provided a careful analysis of the variations in the water level of the Caspian Sea, which he believed followed a 35-year cycle, with alternating wet/cool and dry/warm periods (Stehr and von Storch, 2000: 8). He later extended his claim to argue that this 35-year cycle took place worldwide: "Climate variations involve changes in temperature, air pressure and rainfall that occur globally at the same time. The length of these variations, i.e., the time that elapses between two extremes of the same kind, is 35 years on an average, sometimes more sometimes less..." (Brückner, 2000: 224).

Brückner's colleague, Julius von Hann, a professor of meteorology, was another prominent figure in climate research at this time. Von Hann was author of the first textbook on climatology and was considered (as Brückner observes in von Hann's obituary) a founder of modern meteorology (Stehr and von Storch, 2000: 14). Reflecting the interest in climate variability at the time, he distinguished between "cyclical" and "progressive" climate changes (Stehr and von Storch, 2000: 14). "Progressive" changes are irreversible variations. "Cyclical", in contrast, implicates not only a temporary character of the changes, but also a periodicity. These "cyclical or non-progressive" variations were considered as being composed of a finite number of "waves" with characteristic periods. Von Hann was sceptical of claims of progressive climate change although he tentatively endorsed Brückner's notion of a 35-year cycle (Edwards, 2013: 65–67).

Brückner emphasised that the cause of the periodicity was unclear: "What causes these characteristic variations of the most basic climatic elements? The ultimate reason is still a complete mystery." (Brückner, 1890: 72). But he was aware that "climate variations deeply affect human life" (Brückner, 1890: 73). As with many climate scientists today he was very interested in the economic, social and political consequences of climate changes. He dealt with issues relating to the influence of climate change on migration, crop yields and trade, as well as on health and shifts in the international balance of power (Stehr and von Storch, 2000: 7).

For example, Brückner believed that variations in rainfall amounts have a direct effect on agricultural production. Figure 3.2 illustrates five yearly averages of rainfall in the eighteenth century. It shows that

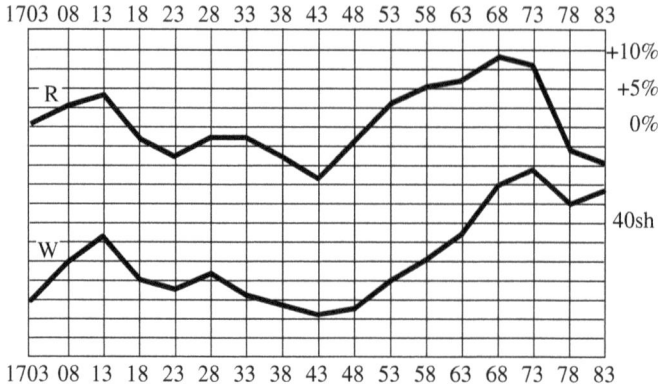

Fig. 3.2. 5 yearly averages of rainfall amounts (R) and wheat prices (W) in England during the 18th century, according to Brückner's analysis. The vertical axis is divided in increments of 2.5 per cent for rainfall and 2 shillings per Imperial Quarter. Redrawn from original. Source: Stehr and von Storch (2000).

precipitation is changing significantly from each five-year segment to the next (variations of plus or minus 5 per cent). Figure 3.2 also shows the price of wheat, which he claimed was closely related to the rainfall amount. He believed that in England's climate, an increase of rainfall would be associated with smaller harvests and thus higher prices. However, he pointed out that while this may have been the case in earlier times, when international commerce was less important, the link would ultimately break down, and he attributed this breakdown to other factors, in particular, political ones.

Brückner further found that in western and middle Europe (with a "maritime climate"), above-average agricultural harvests occurred in warm and dry weather periods. Conversely, a comparable decline in productivity took place in wet and cool atmospheric periods. For both continental Russia and the central United States ("continental climates"), he discovered the inverse; summer rain would be favourable for agriculture. He argued that this geographical pattern of climate changes affected migration from Europe to the United States and he believed his argument was supported by emigration numbers and precipitation statistics — "the stream of immigrants to the United

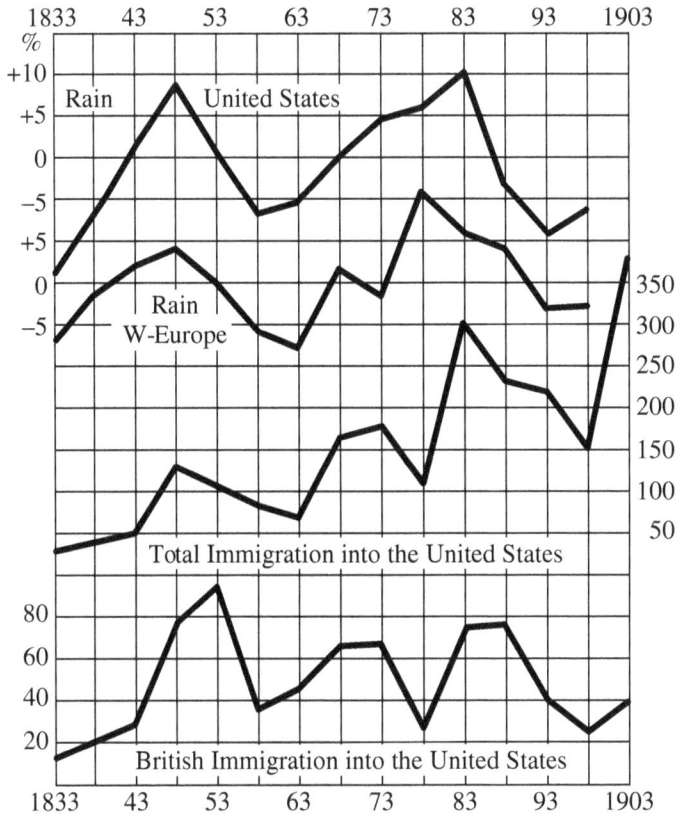

Fig. 3.3. Brückner's analysis of rainfall variations and emigration to the United States in the nineteenth century.

States ebbs and flows with the oscillations of climate, which give it a rhythmical impulse" (Brückner, 2000: 311) (see Figure 3.3).

Clearly, despite his acknowledgement that there may have been "other causes" to patterns of migration, there were many lacunae and problems in Brückner's analysis. Indeed, the idea of climate cycles was later discarded. More recently there has been a renewed interest in climate cycles (Berger *et al.*, 2002). Nonetheless, contemporary climate science is predominantly concerned with, to use von Hann's terminology, not "cyclical" but "progressive" change. It is worthwhile to remind ourselves that climate change itself is not new, and neither is research into its causes and consequences. A hundred years ago, just like today, scientists and policymakers with differing backgrounds and perceptions were engaged in these discussions.

The intense scientific discussion of climate variability at the turn of the twentieth century quickly disappeared from the scientific and public agenda. A new consensus, dominating until the 1970s, almost completely wiped climate change from scientific consideration; it was instead conventionally maintained that climate variations were only episodic in nature, and because of the inherent climatic equilibrium, small and insignificant with respect to impact. A prominent scientific topic can quickly become a marginal one (Stehr and von Storch, 2000: 21). In the last 50 years, climate change has moved again into the forefront of scientific research and social debate. Next, then, we turn to consider the contemporary debate of climate change, and its natural and social causes and consequences.

3.2 Natural Climate Variability

Consider the changes in global temperature measured over the period 1880–2017. Figure 3.4 shows the temperature anomalies, from a

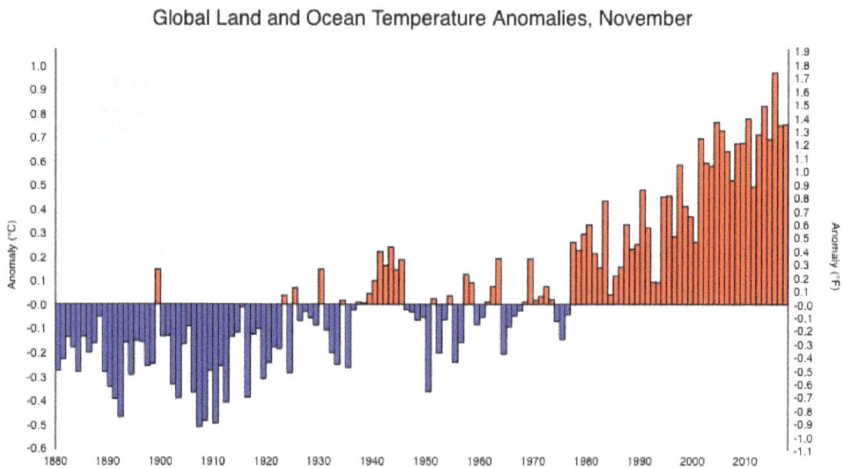

Fig. 3.4.   NOAA National Centers for Environmental Information, Climate at a Glance: Global Time Series, published December 2017, retrieved on 19 December 2017, http://www.ncdc.noaa.gov/cag/.

twentieth century baseline average. It is evident that over the last 140 years the global average has become significantly warmer. According to the National Oceanic and Atmospheric Administration (NOAA), who provided this data, the average global temperature (for both land and ocean surfaces combined) for the months of November in 2016 and 2017 tied as the joint fifth highest on record since 1880 (NOAA, 2017).

These changes are fully consistent with the explanation of anthropogenic climate change, as we consider in the next section. But how can scientists know for sure if such changes are "cyclical" rather than "progressive", and if they are "anthropogenic" rather than "natural"? One of the biggest questions in climate change research is how to distinguish between changes that arise from human action, and those that are purely natural phenomena (Pielke Sr. *et al.*, 2009). In order to research this question, scientists need data from as long a period as possible (Rodriguez *et al.*, 1993: 5). Because of missing or inhomogeneous observation data, climate variations are generally insufficiently documented and understood.

Nonetheless, it is established that climate variability occurs with and without human forcing, and at different geographical and temporal scales. When we speak of "variations", we refer to deviations from a "normal condition". These deviations persist for different lengths of time and will alternate with variations in the opposite direction. But the positive and negative deviations do not equalise in the long run. Thus, the "normal condition" is actually no more than an imaginary value, because there is no "normal" in the geological history of Earth. "Normal" and "deviations," or as they are sometimes called "anomalies", are mathematical constructs. The World Meteorological Organisation has set the averaging interval to be 30 years. This length of 30 years is not a natural constant, but a socially agreed convention that matches with the time horizon of human experience. It is rumoured, in fact, that the specific choice of 30 years is related to the quasi-period of 30–35-year cycles proposed by Brückner.

One of the most studied and large-scale oscillation in climate is the tropical El Niño Southern Oscillation (ENSO). This is an irregular phenomenon that occurs roughly between every 2 to 7 years. Observed at the end of the nineteenth century by H. H. Hildebrandson (1897), this phenomenon is now understood as involving *teleconnections* — links in climate phenomena between seemingly distant parts of the earth. It occurs as a result of the interactions between atmospheric and oceanic processes in the Pacific. The results of ENSO are marked precipitation anomalies; warm phases of the oscillation (El Niño) tend to suppress Atlantic hurricane activity and cool phases (La Niña) tend to enhance it. It also changes the pattern of tropical cyclone activity and droughts in Australia.

Another example of a large-scale climate anomaly is the North Atlantic Oscillation, which describes an anti-correlation of atmospheric pressure and the temperature in the area of the North Atlantic. If the temperatures over Greenland are higher than normal, then as a rule, they are lower than normal over Northern Europe, and vice versa. Coupled with that "seesaw" is a raised atmospheric pressure over Iceland and a reduced atmospheric pressure over the Azores, and vice versa.[3]

Natural climate variability can be seen on far longer time scales, in the occurrence of ice ages, studied through the analysis of ice cores. The "Little Ice Age" in northern Europe from about 1500 to 1750 is an example of a climate anomaly lasting for centuries. Another is the "Younger Dryas" event that started about 12,800 years ago, during which there was an abrupt return to near glacial conditions in northern Europe lasting for about 1,200 years.[4] The Younger Dryas period also ended extremely suddenly — it is estimated that the average annual temperature increased as much as 10°C in ten years. However, the reasons for these abrupt changes are still not known (Lamont-Doherty Earth Observatory, 2003).

---

[3] This mechanism was possibly described for the first time by the Danish missionary Hans Egede Saabye in the eighteenth century.

[4] The name of this event arises from the fact that it is the most recent period in which the *Dryas Octopetala*, a plant characteristic of a cold climate, was found in Scandinavia. See Lamont-Doherty Earth Observatory (2003).

Climate variations can arise from a number of processes. One important process is the orbit of the planet. Serbian astronomer Milutin Milankovic described in 1930 in what is now known as the "Milankovic cycles", in which ice ages could be explained by periodic variations in the parameter of earth's orbit (the shape of the earth's orbit around the sun and the tilt of the earth in this orbit). This "Milankovitch" theory, which has according to some scientists now acquired "textbook status" (Berger, 2002: 104), can help to explain many, but not all, aspects of the alternations of glacial and interglacial periods over tens of thousands of years and has greatly advanced geophysical understandings of the climate system (Berger, 2002).

Variation is caused not just by the planet's orbit but also by changes to its topography. For example, mountains block the flow of air, leading to greater precipitation in some regions, and less in others, creating "rain shadows". Precipitation, in turn, has a "geomorphic" effect on the mountain landform, by causing erosion and thus potentially influencing "height, slopes, peak elevations and channel concavities" (Anders, 2016). Volcanic eruptions also have an impact on the climate, although this impact works in both directions. On the one hand a volcano can have a cooling influence by blasting particles, such as ash, into the air and blocking the sun. The eruption of the volcano Tambora in Indonesia in 1815, for example, led to "The Year Without a Summer" in Europe and North America. Although the cause wasn't known for another 100 years, the eruption led to a drop in global temperatures. Climate scientists and climate historians working with a small amount of direct data as well as indirect climate indicators have revealed that 1816 was probably the coldest year in the last 250 years (Brönnimann and Krämer, 2016: 7). The data indicates that the adverse weather situations that led to a reduced harvest were not more intense, but they were more frequent. As Stefan Brönnimann and Daniel Krämer put it in their account, "While weather was not extreme in 1816, climate was." (Brönnimann and Krämer, 2016: 20)

On the other hand, however, the emission of carbon dioxide and methane from volcanoes (both in repose and erupting) can also contribute to global warming, too. Volcanoes are estimated to release, in total, around 100–300 million tonnes of CO_2 each year. This may be a great deal, but it amounts to only around 1 per cent of the CO_2 that humans release from burning fossil fuels. Nonetheless, at certain times in the past, the eruption of volcanoes may have significantly, albeit temporarily, had a warming effect (Hards, 2005).

Solar activity may also affect the climate. The possibility that sunspot cycles have a periodic influence on the earth's climate has historically played an important role in considerations regarding climate variations. Sunspots are storms that appear as dark marks on the sun. While it might be assumed that dark spots would *reduce* the amount of emitted solar radiation, the opposite correlation is the case: the radiation from the sun comes from its poles and from the "faculae" (the bright rings surrounding the sunspots), thus it is actually the times when sunspots are abundant that the amount of solar radiation arriving at Earth is at a maximum (Ruddiman, 2008: 304). However, the role of sunspots in climate variation seems negligible (Ruddiman, 2008: 305), although the research on this topic is ongoing (IPCC, 2013).

As components of a highly complex, interconnected climate system, it is not possible to know the exact extent to which of these "natural" processes may be affecting the climate. Into this picture of variable climate cuts human activity. It is clear that industrialisation, urbanisation, deforestation and other such social processes have impacted the climate, exacerbating the unpredictability of climate risks and hazards. It is this human-forced climate change that we turn to next.

3.3 Anthropogenic Climate Change

At the end of the eighteenth century, Johann Gottfried Herder described the influence of human activity on the climate in compelling language:

> "Climate is a compound of powers and influences, to which both plants and animals contribute.... Since he stole the fire from heaven and rendered steel obedient to his hand; since he has made not only beasts, but his fellow men also, subservient to his will, and trained both them and plants to his purposes; he has contributed to the

alteration of climate in various ways. Once Europe was a dank forest; and other regions, at present well-cultivated, were the same ... We may consider mankind, therefore, as a band of bold though diminutive giants, gradually descending from the mountains, to subjugate the earth, and change climates with their feeble arms. How far they are capable of going in this respect future will show." (Herder, 1800: 176)

These "bold diminutive giants" have long disrupted the climate, through the use of their various technologies that affect the land, sea and air. As Herder notices, the future impact of this is highly unpredictable. But Herder would not have been aware, of course, of the nature of the chemical processes that partly constitute societies contribution to climate variation.

As we saw in Chapter Two, the temperature of the earth's surface is affected by the ability of the earth's atmosphere to absorb long-wave radiation, and this depends on the chemical composition of the atmosphere. Higher concentrations of absorbing substances in the atmosphere therefore lead to higher atmospheric temperatures. These substances are known as "greenhouses gases" and include water vapour, carbon dioxide, chlorofluorocarbon and methane. A certain portion of "radiatively active" gases in the earth's atmosphere is necessary in order to make temperature that is hospitable to life possible. It is as absurd to speak of carbon dioxide as a "poison", as it is to suggest that "we call it life."[5]

The composition of the atmosphere has indubitably changed through time. Today, however, the concentration of the radiatively active gases is dramatically raised, primarily by the burning of fossil fuels by human societies. Industry, transportation and an exponential growth in the world's population have all led to an increased use of coal, oil and gas, the burning of which has released carbon as a gas. The deforestation that began 8,000 years ago in the Stone Age

---

[5] The Competitive Enterprise Institute (CEI) (funded by Exxon Mobile and the American Petroleum Institute) launched a costly series of TV advertising spots titled "We call it life" aimed at rebuffing the campaign for action on anthropogenic climate change.

was an early influence of human society on the climate. In what is now Europe, China and India, human beings with flint axes created clearings for growing crops, and either burnt the trees or left them to rot, releasing carbon dioxide into the atmosphere (Ruddiman, 2008: 285). Since then, atmospheric carbon dioxide concentration has risen to level unprecedented in at least the last 800,000 years (IPCC, 2014: 4). Methane emissions into the atmosphere have sharply risen too. Methane is emitted from rice fields and from domestic animals like cows, but also from the manufacture and transport of natural gas. Figure 3.5 shows the changes in atmospheric greenhouse gas concentrations of carbon dioxide, methane and nitrous oxide, another greenhouse gas. The heating effect that results from these anthropogenic emissions has led to an "enhanced Greenhouse Effect" (Zillman and Sherwood, 2017).

Globally averaged greenhouse gas concentrations

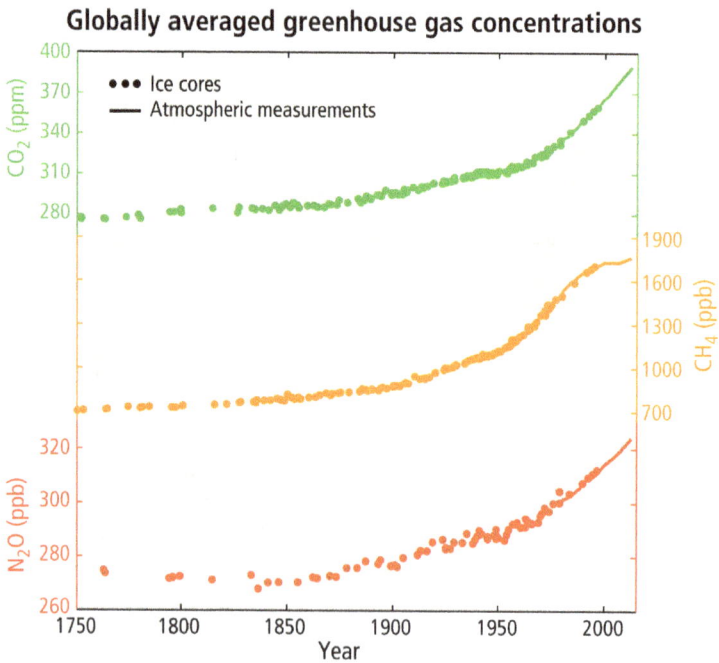

Fig. 3.5. Observed changes in atmospheric greenhouse gas concentrations. Atmospheric concentrations of carbon dioxide (CO_2, green), methane (CH_4, orange) and nitrous oxide (N_2O, red). Data from ice cores (symbols) and direct atmospheric measurements (lines) are overlaid. Source: IPCC (2014).

Exacerbating a rise in greenhouse gas emissions are the concomitant changes in the earth's albedo, from changes in land use, such as deforestation. Since the surface characteristics of the earth influence the thermal emission as well as the vertical transport of heat and moisture in the atmosphere, changes in surface characteristics modify the discharge of long-wave radiation.

At this point we can return to the example of Greenland. As we explained above, Greenland has seen significant climatic variability in the past. However, climate anomalies have drawn scientific attention recently. Between 2002 and 2017, the ice sheet that covers 80 per cent of the island has melted at an average rate of between 264 and 279 gigatons a year (Tedesco *et al.*, 2017), see Figure 3.6. The ice sheet regularly melts in summer, but in recent years the melting has started earlier; average surface temperature and the number of "melt days"

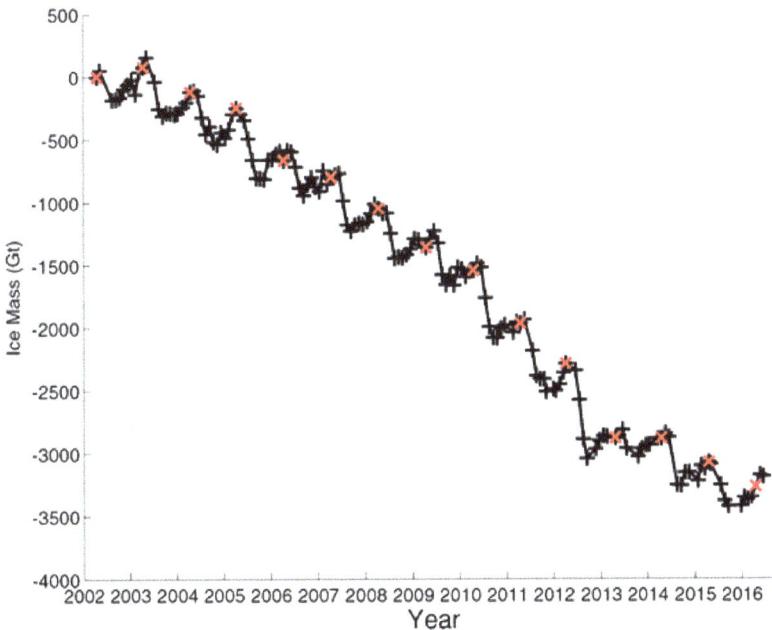

Fig. 3.6. Monthly change in the total mass (in gigatons) of the Greenland ice sheet between April 2002 and April 2016, estimated from GRACE measurements. The red crosses denote the values for the month of April of each year. Source: Tedesco *et al.* (2016).

have risen. The degree of melting and its contribution to sea level rise (prior to 2002) remains contentious because of a lack of data (Jevrejeva, 2016). Yet some scientists regard this issue as significant, because this ice sheet holds (in volume) the equivalent to 7 metres of sea level rise (Klintisch 2017). It is an important cog in the global climate system (Tedesco et al., 2017). The ice sheet contributes to the "albedo affect" by reflecting the sun's energy back into space. The albedo is being reduced by the increased amounts of microbes and algae that grow on the wet surface of the melting ice sheet which boosts its absorption of solar energy. This leads to a feedback loop, driving further melting (Klintisch, 2017). And the melting glaciers not only indicate rising global temperatures, they also contribute to them; a dropping albedo coincides with a release of carbon and methane into the atmosphere, as the organic matter that is contained in the thawing permafrost decays. If the thawing continues, it is estimated that thawing by permafrost will emit the equivalent of 850–1,400 billion tonnes of CO_2, thus further driving global warming.

3.4 Measuring and Modelling Climate Change

There are various procedures for studying climate variability. One approach is to analyse observation data. The data in Figure 3.6 comes from a pair of satellites called the Gravity Recovery and Climate Experiment (GRACE). These satellites were launched in 2002 to make monthly observations of changes in the earth's gravity, providing estimates of mass loss from ice sheets in Greenland and Antarctica and glaciers, and tracking mass movement of water (Jevrejeva et al., 2017). But there is only 16-years-worth of data, a short time for understanding climate variation. This makes it difficult to reliably assess large-scale and long-term conditions.

There have been global observations for only about 150 years, and this supposedly global set of data has large spatial gaps. Large areas of the Pacific and the Southern Ocean were for many years hardly traversed by ships, so there is hardly any data for these regions. Good sets of data with greater spatial resolution and quality-assured observations have existed for only perhaps 30 years, ever since satellites were routinely employed. This data is, of course, inadequate for a description of climate variations extending decades, let alone a longer time-period.

Besides instrumental data routinely collected over the past ~150 years by meteorological and oceanographic services, there is also "indirect" data. Ice cores, for example, offer an important source of information about past climate. Ice cores are cylinders drilled out from an ice sheet or glacier, from depths of up to 3 kilometres. The ice contains small bubbles that provide a sample of the atmosphere, from which it is possible to measure past temperature and concentration of gases, up to a period of hundreds of thousands of years. The oldest ice core, drilled by the European Project for Ice Coring in Antarctica (EPICA), extends back 800,000 years and shows a succession of long, cold "glacial" periods, interspersed roughly every 100,000 years by warm "interglacia" periods (of which the last 11,000 years is the most recent) (British Antarctic Survey, 2015). The apparent parallel development of temperature and CO_2 concentrations is also evident — warm temperatures accompany elevated CO_2 concentrations, and vice versa (see Figure 3.7). As one report puts it: "We see no examples in the ice core record of a major increase in CO_2 that was not accompanied by an increase in temperature" (British Antarctic Survey, 2015).

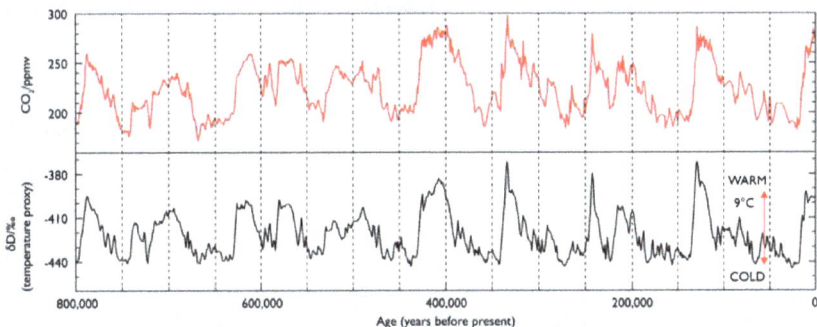

Fig. 3.7. Ice core data from the EPICA Dome C (Antarctica) ice core: deuterium (D) is a proxy for local temperature; CO_2 from the ice core air. Source: British Antarctic Survey Science Briefing (2015).

The width of tree rings, or the deposit characteristics of deep-sea sediments, or isotope relationships in limestone shells in ocean sediment also yield information about past climate variations. Interpreted by experts, these pieces of information provide a wealth of information about climate variations in the course of hundreds, thousands, or even millions of years (Crowley and North, 1991).

In addition to interpreting data, scientists studying climate change utilise detailed "quasi-realistic" models of the climate system (as discussed in Chapter Two). Climate models are used to project possible climatic futures. These models, while increasingly sophisticated, do not and cannot, produce watertight predictions. Instead, they usually offer a range of plausible pathways, or future "scenarios". A "scenario" incorporates data on various components of the climate system and the relationships between them: greenhouse gas emissions, atmospheric concentrations and temperatures are incorporated alongside other "natural" climate variations, as well as with energy policies and technological change (Hayhoe et al., 2017).

Social trends clearly play a key role in determining the amount of emissions. Each scenario is based upon particular "storylines" that lay out "a consistent picture of demographics, international trade, flow of information and technology, and other social, technological, and economic characteristics of future worlds" (Hayhoe et al., 2017: 135). These storylines are used to derive a certain level of greenhouse gas emissions, which are fed into a climate model. There are between 30 and 40 prominent climate models providing different projections of global warming for a given increase in greenhouse gas concentrations (Brown and Caldeira, 2017). Different models project different amounts of warming, primarily because no consensus exists on how best to model the various aspects of the climate system. Comparative analysis of climate change models suggests that models should be combined with observations to reduce the degree of uncertainty (Brown and Caldeira, 2017; Müller and von Storch, 2004).

The growing sophistication and detail of these models notwithstanding, it is clearly impossible to include all potential effects and non-linear developments in a complex, chaotic interactive system consisting not only of "natural" processes but social activities too (Trenberth, 1997: 131). The social component not only increases the complexity of the system but also adds an element of human reflection, intentionality and (non-rational) behaviour. Correctly projecting the state of the future global climate would require accurately predicting the complex actions and interactions of human beings in the next century. And, of course, since human beings respond to these predictions *themselves* in unpredictable ways, the problem is sharpened (Ruser, 2014: 173–174).

A scenario then is not as feeble or fatuous as a tealeaf reading, but neither is it as precise as a weather forecast. Ultimately, uncertainty remains. While models help us to understand climate variability and grasp possible futures, they cannot provide definite predictions, however much policymakers might like them to (Machin and Ruser, 2018). But if climate change is difficult to measure and impossible to predict, then where does this leave the policymakers who wish to become informed on the science and the scientists who wish to inform them?

We turn here to consider the role of the Intergovernmental Panel on Climate Change (IPCC). The IPCC is the largest, most renowned and widely cited scientific body in the world regarding climate change. Created by the United Nations Environment Programme and the World Meteorological Organisation in 1988, the IPCC does not itself conduct research. It reviews and assesses relevant scientific, technical and socio-economic information regarding climate change. The fifth assessment report became available in 2014.

The first part of the fifth assessment report is the Physical Science Basis report, in which peer-reviewed contributions from 259 authors from thirty-nine countries are compiled. This report explains that it is clear that anthropogenic climate change is occurring. It states that "warming of the climate system is unequivocal" and that the "unprecedented" changes in the system over the last sixty years are "extremely likely" to have been influenced by human activity (IPCC, 2013). It explains that atmosphere and ocean temperatures have *already* increased, that the ice sheets are *already* diminishing and that sea levels have already risen. This is due to the rise in concentrations of greenhouse gases and the report confirms that even if emissions are reduced today, climate change will persist for centuries. There is "high confidence" that the Greenland ice sheet will be entirely lost (producing, as we mention above, an average sea level rise of up to seven metres). "Continued emissions of greenhouse gases will cause further warming and changes in all components of the climate system. Limiting climate change will require substantial and sustained reductions of greenhouse gas emissions," the report states bluntly (IPCC, 2014: 8).

In order to make future projections, the IPCC utilises climate modelling to make future projections. In contrast to earlier sets of scenarios (which were based on emissions), the scenarios used here, known as Representative Concentration Pathways (RCPs), are based on radiative forcing. Radiative forcing is the measurement of the capacity of a gas to affect the energy balance of the atmosphere. This provides "a simple yet fundamental index that allows us to look at how climate change is driven by the energy imbalance of the Earth System" (National Research Council, 2005: vii). The four RCPs are numbered according to the change in radiative forcing by 2100, relative to pre-industrial levels (+2.6, +4.5, +6.0 and +8.5 watts per square metre). In one RCP, greenhouse gas emissions have been reduced, in two they have been stabilised and in RCP8.5 they remain high. Only under the lowest scenario (RCP2.6) is global temperature likely to remain under 2°C. Under the highest scenario (RCP8.5) there is an increase of 5°C to 8°C above preindustrial temperatures; in this scenario Greenland and Antarctic ice sheets could melt entirely.

These scenarios are useful ways of presenting possible futures to policymakers. The IPCC Synthesis Report Summary for Policymakers explains that by the middle of this century, the extent of climate change is substantially affected by the choice of scenario (IPCC, 2014: 10). Yet the ability of these models to simulate fine-scale processes at smaller scales is limited (Hayhoe *et al.*, 2017: 1341). The further into the future projections are made, the greater the degree of uncertainty (Hayhoe *et al.*, 2017: 139).

The role the IPCC plays in translating between scientists and policymakers is not easy and it has been widely criticised — for example, IPCC has been accused of pushing a particular political agenda. However, the IPCC does make it clear that none of its predictions are regarded as certain, and that there is no one "right" way to combat climate change. And yet the very fact that the issue is framed as one single problem best grasped at the global level may undermine a more sensitively drawn account (Machin, 2013). The maps in the IPCC reports show regional variation and it mentions the different constraints and capacities of different parts of the world, but it summarises all its data by pulling it back to world averages. The concern is that this glosses over the differentiated affects and responses at a local level, as we consider next.

3.5 The Impact and Inequality of Climate Change

Knowledge of the causes and consequences of climate change is inevitably imperfect for the climate system is highly complex and involves the analysis of data from a myriad of sources; different processes interconnect in different ways and have different tipping points. Accurately predicting the future climate is, as we have seen, impossible. But the possible impact of climate change on human lives is an even more complicated issue — not only is climate change likely to have mixed economic, political and social consequences (i.e., affecting agriculture, living standards and human health), but the way in which societies respond to extreme weather and changing climate will abate or sharpen these consequences. There may be a reaction to both changing climate and the predictions of a changing climate: new learning processes may be set in motion, technological development and implementation may be instigated, and policies and social movements may underpin a change in lifestyle patterns. These responses will overlay the effects considerably, although not evenly. It is possible that these responses — to both the risks and realities of a changing climate — will affect the climate even further, creating a complex feedback loop in an already complex system. In this section, we discuss the anticipated direct consequences of climate change and the risks it poses for human societies.

In studying climate change impact, some of the most important questions concern the time and location of expected climate changes. A frequently used characteristic measure of anthropogenic climate changes is the *global* average air temperature. This quantity functions meaningfully as a global indicator of the intensity of climate changes, but it is practically irrelevant when we are trying to understand and assess *regional* impacts, and the actual meaningful implications for both society and ecology. For in some regions the temperature will rise more quickly than in others; in a few areas it may even cool down. Globally, an intensification of the water cycle of evaporation and precipitation can be expected, so that more rainfall seems possible globally. Again, however, this indicator is largely meaningless for societies, since the precipitation distribution could also be shifted, so that a few areas will experience more rain and others less.

A similar issue arises with the important and commonly discussed aspect of climate change — its impact on sea levels. The thermal expansion of seawater combined with the destabilisation of ice sheets in Antarctic and Greenland and land subsidence may produce a rise in sea levels of between 30 cm and 2 metres by 2100 (Treuer *et al.*, 2018: 108). The global rate of rise is currently 3–4 mm per year, but this rate appears to be accelerating and could increase exponentially (Treuer *et al.*, 2018: 109; Vitousek *et al.*, 2017). Although sea level rise is smaller than normal ocean-level fluctuation, it still increases the risk of frequency and severity of coastal flooding, which are difficult to predict (Vitousek *et al.*, 2017). Furthermore, this risk will not be uniform across all regions (Jayanthi *et al.*, 2018: 1)

Coastal communities and low-lying areas are particularly at risk, but vulnerabilities differ, due to differences in geology and socio-economic resources. A region that is particularly vulnerable to sea level rise is South Florida. Not only is much of this region low-lying, it also sits upon porous limestone bedrock that limits the effectiveness of levee and pump systems. Without adaptation, human displacement and economic losses are expected to be enormous. The degree of vulnerability, however, can be reduced by adaptation measures. The city of Miami Beach is currently investing $500 million into flood defences. Yet, as Treuer and colleagues point out, "As a relatively small and wealthy community … Miami Beach is the exception in terms of its active response to sea level rise and its financial capabilities to do so." (Treuer *et al.*, 2018: 109). Poorer regions do not have such capacity.

The Nagapattinam district in Tamil Nadu is also a low-lying region on the coast of South East of India that is particularly at risk. Sea level rise is likely to affect forest, wetlands agriculture and therefore also the biodiversity of the region and livelihoods of the population there (Jayanthi *et al.*, 2018: 11). Adaptation is needed, for example, barriers and buffers along the coast, salt-tolerant crops, a changing crop calendar, insurance schemes, and so on. But constraints implementing such adaptions stem from "limited financial resources, lack of transparency in implementation and monitoring, absence of public participation and lack of integrated coastal zone management and different opinions of risks" (Jayanthi *et al.*, 2018: 12). Not only do some regions have more resources than others to combat sea level rise, they are also more willing to utilise them (Treuer *et al.*, 2018).

The island states and territories in the Central and South Pacific Ocean are also extremely vulnerable to rising sea levels, which threaten homes, livelihoods, human health and food security (Germanwatch, 2004; Barnett, 2005). One of these island-states, Tuvalu, which lies in its entirety less than three metres above sea-level, has been named a "titanic state" and has become a particularly prominent symbol for climate change. Here "a change in sea level is therefore not an abstract risk but a challenging task to the everyday life of Tuvaluans" (Germanwatch, 2004: 6). Tuvaluans have been labelled the world's first "climate refugees". This is not merely a media slogan but has had socio-political repercussions. In 2002, New Zealand launched the Pacific Access Category (PAC) programme, in which 75 citizens from Tuvalua, along with 75 from Kiribati, 250 from Tonga and 250 from Fiji are granted residence each year. Not all individuals qualify, however, as applicants must be in good health, be of good character, speak English, have a job offer in New Zealand, be able to support themselves and their family, and be between 18 and 45 years of age. The statement on the New Zealand immigration fact sheet explains that a climate refugee will be refused entry if "you don't meet our character requirements", which perhaps most succinctly reveals the power relations at work in the politics of climate change.[6] Wealthy nations ultimately control the movement of climate refugees "whose plight is a product of Western lifestyles" (Jerolmack, 2015). A critical analysis of the "climate refugee" label shows that the portrayal of the inhabitants of Pacific Islands as passive victims somewhat overlooks the complexity of the problem and the possibility of risk reduction (Barnett, 2005: 216; Farbotko and Lazrus, 2012).

How far future technical innovations and practices will make new adaptation strategies possible, even in wealthy nations, is impossible to predict. Adaptation in agriculture, by changing management concepts or by choosing different crops, which may also be bred to meet certain climatological and other conditions, can be carried out in a time frame of a few years, which is a short time scale compared to the speed of climate change. It may be that agriculture in some parts of the world may even benefit from changes in the climate.

The anticipated climate changes could have a multitude of direct and indirect health consequences. As a report released by the Worldwatch Institute and the United Nation Foundations states: "We are ... profoundly changing our planet's climate. It is increasingly apparent that the breadth and depth of the changes we are wreaking are imperilling not only many other species, but the health and wellbeing of our own species as well." (Myers, 2009: 2). The various public health threats identified by the report include: infectious diseases, food scarcity, water scarcity, air pollution, natural disasters and population displacement (Myers, 2009: 2). The report suggests that these threats together constitute the greatest public health challenge of the twenty-first century (Myers, 2009: 2).

---

[6] See visa conditions at: www.immigration.govt.nz/new-zealand-visas/apply-for-a-visa/visa-factsheet/pacific-access-category-resident-visa#conditions

The World Health Organisation (WHO) also warns that global climate change is likely to damage agriculture and health-infrastructure, as well as lead to a significant rise in illnesses and contagious diseases. It notes that 250,000 additional deaths per year should be expected from malnutrition, malaria, diarrhoea and heat stress, all connected to a changing climate (WHO, 2017). In a WHO report on the health impacts of climate change on small island states, particular diseases are identified that are climate-sensitive: malaria, dengue, diarrhoeal disease, heat stress, skin diseases, acute respiratory infections and asthma (Ebi *et al.*, 2005: 20). It explains that the incidence of malaria, for example, has increased and is correlated to a rise in atmospheric humidity and temperature, although this correlation is not simple (Ebi *et al.*, 2005: 24).

These reports also highlight that some populations in some parts of the world are more vulnerable than others. They also note the role of a robust policy response aimed at securing "resilience" of these populations. This leads us to a crucial observation. Paul Reiter (2001) the former chief entomologist at the US government's dengue research lab in Puerto Rico, points out that even though malaria is thought of as a "tropical disease", it was until recently widespread in both Europe and North America:

"In the 1880s, virtually all the US was malarious, and even parts of Canada. When ... the Center for Disease Control and Prevention (CDC) was founded in 1946, its principal mission was to eradicate malaria from the US. In Europe, the disease was endemic as far north as Norway, Sweden and Finland. In the 1920s, epidemics killed hundreds of thousands in the Soviet Union, right up to the Arctic Circle. One of the last European countries to be freed of the disease was Holland. That was in 1970. As for dengue, the principal vector has been living happily in North America for about 300 years..."

Reiter argues that the natural climate would only be a minor factor limiting or favouring malaria. Instead, the organisation of society and the precautionary health measures taken would be far more relevant.

Indeed, there is a bigger point here. The debate about the impact of climate change is also always a debate about the relative importance of environmental conditions versus the importance of social organisation. Socio-economic factors are arguably far more significant than climate in the determination of disease prevalence. For example, there are *one thousand times* more cases of dengue in the Northern regions of Mexico than in Southern Texas. The climate across this 100-kilometre band is practically the same, even the vector habitats are similar in many instances, but the pattern of social interactions and access to public health are vastly different. Socialising outside at dusk, when mosquitoes are most active, is prevalent practice in Mexico. North of the border, people are more often indoors in air-conditioned rooms. Air conditioning is an adaptive measure, which can protect people from vector-borne diseases as well as heat waves.

Paradoxically, of course, air conditioning uses up a great deal of energy. In the United States, air conditioners use about 6 per cent of all electricity produced in the country, resulting in the release of 117 million metric tons of carbon dioxide into the atmosphere.[7] This example thus neatly exemplifies the pattern highlighted by scholars of climate justice: those benefitting from the lifestyles that have heightened climate change are those who are most equipped to protect themselves from its worst affects.

Work in political ecology crucially shows that environmental factors are inevitably entangled with economic and social specificities. Thus, the parts of the world and sectors of society that are the least well-off are more likely to suffer from environmental hazards. For Ulrich Beck, emerging environmental issues smooth out the uneven distribution of earlier social threats: "Hunger is hierarchical ... Nuclear contamination, however, is egalitarian and in that sense 'democratic'. Nitrates in the ground water do not stop at the general director's water tap." (Beck, 1992: 109). According to Beck, we are all equally exposed to the hazards of climate change. And yet, confusingly, he admits later

---

[7] See https://energy.gov/energysaver/air-conditioning

in the same paper that the greenhouse effect will disproportionately affect those who can least afford it: "The poorest in the world will be hit the hardest" by environmental change (Beck, 1992: 110).

It has been widely acknowledged that the regions of the world that will most likely be worst affected by a changing climate are the regions that have least contributed to them. Differences in historical emissions have raised the question of the existence of differentiated climate responsibilities. Theorists point to the unequal world order instituted by colonialism and argue that such socially produced inequality will be sharpened by climate change (Caney, 2012; Neumayer, 2000). It is also true that domestic inequalities *within* states are aggravated by the disproportionate effects of climate change. Research on "heat death", for example, reveals its similar unequal distribution across society. During the 1995 heat wave in the city of Chicago (Klinenberg, 2002; Browning *et al.*, 2006), heat wave mortality was negatively linked with neighbourhood affluence and positively linked to commercial decline. Immigrants, indigenous, black, poor and working-class communities are more likely to suffer the effects of environmental damage. As Juliet Schor explains, in the United States at least, the distribution environmental hazards and environmental amenities are both skewed by race and income (Schor, 2015).

Analysis of Hurricane Katrina that made landfall in August 2005 on the Gulf Coast of the United States attests to the interconnection of environmental risk, structural racism and patterns of economic and political inequality that ensured that the effects of the hurricane were distributed unevenly (Hartman and Squires, 2006; Sharkey, 2007). When the levees protecting the city of New Orleans were breached, 80 per cent of the city was flooded, killing over a thousand people and displacing more than a million in the Gulf Coast region. Hurricane Katrina was the largest residential disaster in the history of the United States (Plyer, 2016), but it particularly affected black neighbourhoods, located in low-lying areas in unsafe buildings (Young, 2006). The same insights apply to more recent devastation in August 2017, wreaked on Texas by Hurrican Harvey. As Ilan Kelman (2017) writes:

"Weather and climate don't cause disasters — vulnerability does … A disaster involving a hurricane cannot happen unless people, infrastructure and communities are vulnerable to it. People become vulnerable if they end up lacking knowledge, wisdom, capabilities, social connections, support or finances to deal with a standard environmental event such as a hurricane."

Gender inequality is also exacerbated by environmental hazards such as climate change (Terry, 2009). Sherilyn MacGregor notices that climate change is not "gender neutral": "Women are more dramatically affected by all forms of environmental degradation than men, due to their social roles as carers and provisioners and in their social location as the poorest and most vulnerable at the bottom of social hierarchy." (MacGregor, 2010: 226). She calls attention, however, to the problematic tendency of constructing women as only vulnerable *victims*: "There is little room for human voices — let alone the voices of those women who would wish to complicate or resist the way they appear in the climate story." (MacGregor, 2010: 227). Climatic crisis reproduces and sharpen existing political, social and economic inequalities. The way they are framed may also reproduce certain social norms.

3.6 Conclusion: Understanding Change

As we hope to have shown in this chapter, climate change variability is not a new research area, but was investigated as far back as the end of the eighteenth century. Climate change research then has precedence, but the extent, seriousness and unpredictability of a changing climate have dramatically increased, alongside its political prominence. The growing width and depth of information and knowledge of the dynamics of climate, has catapulted climate change into the public sphere. Paradoxically, while science powered the emergence of industrial society, dependent upon the mining of fossilised energy resources, it is science that has revealed the impacts of the burning of these resources and indicates its possible repercussions.

These repercussions however will not be evenly spread. We have also highlighted in this chapter the likelihood that Anthropogenic climate change, on the scale that now seems likely, will exacerbate inequality. It is also possible that inequality undermines social cohesion, which can reduce the "socio-ecological resilience of communities" (Laurent, 2013) and their ability to act collectively to combat climate change.

This is why we suggest that the social sciences must contribute to the scientific endeavour of understanding the significance of climate change on society. The impacts of changes in the natural environment and climatic foundations of human life can only be stated with difficulty and uncertainty. Even trickier is the assertion of societal, cultural and political consequences, not only for predicting future dangers, but also of social reactions to those predictions themselves. Of course, we know that there will be impacts, but the difficulty is to determine what they might look like and how we might mitigate the effects of such impacts. It is all too easy to blame "natural catastrophe" for what is *socially* and *politically* generated suffering, inequality and risk. It is particularly important to remember this when faced with the tendency to assume that climate and climate change works to *determine* society and that the impact of climatic phenomena can be measured and predicted without taking into account social factors and human agency. It is to the problematic body of literature on "climate determinism" that we turn to in the next chapter.

References

Adamson, P., Aliaga, *et al.* (2017) "Constraints on oscillation parameters from ν_e appearance and ν_μ disappearance in NovA," *Physical Review Letters* **118**(23): 231801.

Anders, Alison (2016) "Precipitation patterns and topography". In: *Vignettes: Key Concepts in Geomorphology*. Available at: https://serc.carleton.edu/vignettes/collection/25201.html (accessed: 15 December 2017)

Barnett, Jon (2005) "Titanic States? Impacts and Responses to Climate Change in the pacific Islands," *Journal of International Affairs* **59**(1): 203–219.

Beck, Ulrich (1992) "From industrial society to the risk society: Questions of survival, social structure and ecological enlightenment," *Theory, Culture and Society* **9**(1): 97–123.

Berger, Wolfgang H. (2002) "A case for climate cycles: Orbit, sun and moon". In: *Climate Development and History of the North Atlantic Realm*. (Eds.) Wefer, Gerold, Wolfgang H. Berger, Karl-Ernst Behre and Eystein Jansen. Berlin and Heidelberg: Springer.

British Antarctic Survey Science Briefing (2015). "Ice cores and climate change". Available at www.bas.ac.uk/wp-content/uploads/2015/04/ice_cores_and_climate_change_briefing-sep10.pdf (accessed: 22 March 2018)

Brönnimann, Stefan and Daniel Krämer (2016) "Tambora and the 'year without a summer' of 1816: A perspective on earth and human systems science," Geographica Bernensia G90.

Brown, Patrick T. and Ken Caldeira (2017) "Greater future global warming inferred from Earth's recent energy budget," *Nature* **552**(7683): 45.

Browning, Christopher R., Danielle Wallace, Seth L. Feinberg and Kathleen A. Cagney (2006) "Neighbourhood social processes, physical conditions, and disaster-related-mortality: The case of the 1995 Chicago heat wave," *American Sociological Review* **71**(4): 661–678.

Brückner Eduard (2000) [1890, 1st ed.] "Climate Changes since 1700". In: *Eduard Brückner — The Sources and Consequences of Climate Change and Climate Variability in Historical Times*. (Eds.) Stehr, Nico and Hans von Storch. Springer.

Brückner, Eduard (2000) [1915, 1st ed.] "The Settlement of the United States as Controlled by Climate and Climate Oscillations". In: *Eduard Brückner — The Sources and Consequences of Climate Change and Climate Variability in Historical Times*. (Eds.) Stehr, Nico and Hans von Storch. Springer.

Caney, Simon (2012) "Just emissions," *Philosophy & Public Affairs* **40**(4): 255–300.

Chen, J. L., T. Pekker, C. R. Wilson, B. D. Tapley, A. G. Kostianoy, J.-F. Cretaux and E. S. Safarov (2017) "Long-term Caspian Sea level change," *Geophysical Research Letters* **44**(13): 6993–7001.

Crowley, T. J. and G R. North (1991) *Paleoclimatology*. New York: Oxford University Press.

Dugmore, Andrew J., Christian Keller and Thomas H. McGovern (2007) "Norse Greenland settlement: Reflections on climate change, trade, and the contrasting fates of human settlements in the North Atlantic Islands," *Arctic Anthropology* **44**(1): 12–36.

Ebi, Kristie L., Nancy D. Lewis and Carlos F. Corvalán (2005) "Climate variability and change and their health effects in small island states: Information for adaptation planning in the health sector," WHO. Available at: www.who.int/globalchange/publications/climvariab.pdf?ua=1

Edwards, Paul (2013) *A Vast Machine: Computer Models, Climate Data, and the Politics of Global Warming*. MIT Press.

Farbotko, Carol and Heather Lazrus (2012) "The first climate refugees contesting global narratives of climate change in Tuvalu," *Global Environmental Change* **22**(2): 382–390.

Germanwatch (2004) *Climate Change Challenges Tuvalu*. Information Booklet available at: https://germanwatch.org/download/klak/fb-tuv-e.pdf (accessed: 12 January 2018)

Hards, Vicky (2005) Volcanic contributions to the global carbon cycle. *British Geological Survey Occasional Publication* No. 10: 1.

Hartman, Chester W. and Gregory O. Squires (2006) *There is No Such Thing as a Natural Disaster: Race, Class, and Hurricane Katrina*. London: Routledge.

Hayhoe, K., J. Edmonds, R. E. Kopp, A. N. LeGrande, B. M. Sanderson, M. F. Wehner and D. J. Wuebbles (2017) "Climate models, scenarios, and projections". In: *Climate Science Special Report: Fourth National Climate Assessment, Volume I*. (Eds.) Wuebbles, D. J., D. W. Fahey, K. A. Hibbard, D. J. Dokken, B. C. Stewart and T. K. Maycock. US Global Change Research Program, Washington, DC, USA, pp. 133–160.

Herder, Johann Gottfried von (1800) *Outlines of a Philosophy of the History of Man*. (Trans.) T. Churchill. New York: Bergman Publishers.

Hildebranson, H. H. (1897) "Quelque recherches sur les entres d'action de l'atmosphere," *Sven Vetenskaps akad. Handl.* **43**: 606–631.

IPCC (2013) *Climate Change 2013: The Physical Science Basis*. Contribution of Working Group I to the Fifth Assessment Report of the Intergovernmental Panel on Climate Change. (Eds.) Stocker, T. F., D. Qin, G.-K. Plattner, M. Tignor, S. K. Allen, J. Boschung, A. Nauels, Y. Xia, V. Bex and P. M. Midgley. Cambridge and New York: Cambridge University Press.

IPCC (2014) *Climate Change 2014 Synthesis Report: Summary for Policymakers*. Contribution of Working Groups I, II and III to the Fifth Assessment Report of the Intergovernmental Panel on Climate Change. (Eds.) Pachauri, R. K. and L. A. Meyer. Geneva, Switzerland: IPCC.

Jayanthi, Marappan, Selvasekar Thirumurthy, Muthusamy Samynathan, Muthusamy Duraisamy, Moturi Muralidhar, Jangam Ashokkumar and Koyadan Kizhakkedath Vijayan (2018) "Shoreline change and potential sea level rise impacts in a climate hazardous location in southeast coast of India," *Environmental Monitoring and Assessment.* **190**(51): 1–15.

Jerolmack, Colin (2015) "Choking on poverty: Inequality and environmental suffering". In: *Public Books*. Available at http://www.publicbooks.org/nonfiction/choking-on-poverty-inequality-and-environmental-suffering

Jevrejeva, S., A. Matthews and A. Slangen (2017) "The twentieth-century sea level budget: Recent progress and challenges," *Surveys in Geophysics* **38**(1): 295–307.

Kelman, Ilan (2017) "Don't blame climate change for the Hurricane Harvey disaster — blame society," *The Conversation*. 29 August.

Klinenberg, Eric (2002) *Heat Wave: A Social Autopsy of Disaster in Chicago*. Chicago, Illionois: University of Chicago Press.

Klintisch, Eli (2016) "Why did Greenland's vikings disappear?" *Science* 10 November. doi:10.1126/science.aal0363.

Klintisch, Eli (2017) "The great Greenland meltdown," *Science* 23 February. doi:10.1126/science.aal0810.

Lamont-Doherty Earth Observatory (2003) "Abrupt climate change," Available at: http://ocp.ldeo.columbia.edu/res/div/ocp/arch/examples.shtml (accessed: 12 March 2018)

Laurent, Eloi (2013) "Inequality as pollution, pollution as inequality: The social-ecological nexus," Working Paper of the Stanford Center on Poverty and Inequality. Available at http://web.stanford.edu/group/scspi/_media/working_papers/laurent_inequality-pollution.pdf (accessed: 8 December 2015).

Lehmann, Philipp N. (2015) "Whither climatology? Brückner's climate oscillations, data debates, and dynamic climatology," *History of Meteorology* **7**: 49–70.

MacGregor, Sherilyn (2010) "'Gender and climate change': From impacts to discourses," *Journal of the Indian Ocean Region* **6**(2): 223–238.

Machin, Amanda (2013) "Climate change is not a fairytale," *Warscapes*. Available at: http://www.warscapes.com/opinion/climate-change-not-fairy-tale

Machin, Amanda and Alexander Ruser (2018) "What counts in the politics of climate change? Science, scepticism and emblematic numbers". In: *Science, Numbers and Politics*. (Ed.) Markus Prutsch. Palgrave MacMillan.

Mahony, Martin and Georgina Endfield (2018) "Climate and colonialism," *WIREs Climate Change*. **9**(2).

Mauelshagen, Franz (2018) "Climate as a scientific paradigm — Early history of climatology to 1800". In: *The Palgrave Handbook of Climate History*.

(Eds.) Sam White, Pfister, Christian and Mauelshagen, Franz. Basingstoke: Palgrave Macmillan.

Mekhtiev, Arif Sh. and A. K. Gul (1996) "Ecological problems of the Caspian Sea and perspectives on possible solutions". In: *Scientific, Environmental, and Political Issues in the Circum-Caspian Region*. (Eds.) M. H. Glantz and Zonn, Igor S. Springer.

Müller, Peter K. and Hans von Storch (2004) *Computer Modelling in Atmospheric and Oceanic Sciences: Building Knowledge*. Springer Science & Business Media.

Myers, Samuel S. (2009) Worldwatch Report #181: Global Environmental Change: The Threat to Human Health. Available at: www.worldwatch.org/bookstore/publication/worldwatch-report-181-global-environmental-change-threat-human-health (accessed: 12 December 2017)

National Research Council (2005) "Radiative forcing of climate change: Expanding the concept and addressing uncertainties," Washington DC: National Academies Press.

Neumayer, Eric (2000) "In defense of historical accountability for greenhouse gas emissions," *Ecological Economics* **33**(2): 185–192.

NOAA (2017) "State of the climate: Global climate report for November 2017". Available at: https://www.ncdc.noaa.gov/sotc/global/201711 (accessed: 19 December 2017)

Pielke Sr., Roger, *et al.* (2009) "Climate change: The need to consider human forcings besides greenhouse gases," *Eos* **90**(45): 413–414.

Plyer, Allison (2016) "Facts for features: Katrina impact," *The Data Centre*. Available at: www.datacenterresearch.org/data-resources/katrina/facts-for-impact/ (accessed: 22 March 2018).

Reiter, Paul (2001) "Climate change and mosquito-borne disease," *Environmental Health Perspectives* **109**(2001): 141–161.

Rodriguez, Roberto, M. Carmen Llasat and Emilio Rojas (1993) "Evaluation of climatic change through harmonic analysis". In: *Recent Studies in Geophysical Hazards*. (Eds.) El-Sabh, M. I., T. S. Murty, S. Venkatesh, F. Siccardi and K. Andah. Springer.

Ruddiman, William F. (2008) *Earth's Climate: Past and Future*. New York and Basingstoke: W. H. Freeman and Company.

Ruser, Alexander (2014) "Sociological quasi-labs: The case for deductive scenario development," *Current Sociology* **63**(2): 170–181.

Schor, Juliet (2015) "Climate, inequality, and the need for reframing climate policy," *Review of Radical Political Economics* **47**(4): 525–536.

Sharkey, Patrick (2007), "Survival and death in New Orleans: An empirical look at the human impact of Katrina," *Journal of Black Studies* **37**(4): 482–501.

Steen, Harold K. (1991) *The Beginning of the National Forest System.* Washington, DC: US Department of Agriculture Forest Service.

Stehr, Nico and Hans von Storch (2000) "Eduard Brückner's ideas — Relevant in his time and today". In: *Eduard Brückner — The Sources and Consequences of Climate Change and Climate Variability in Historical Times.* (Eds.) Stehr, Nico and Hans von Storch. Springer.

Tedesco, M., *et al.* (2017) "Greenland ice sheet". In: *Arctic Report Card 2017.* http://www.artic.noaa.gov/Report-Card (accessed: 5 Janaury 2018)

Tedesco, M., *et al.* (2016) "Greenland Ice Sheet". In: *Arctic Report Card 2016.* http://www.artic.noaa.gov/Report-Card (accessed: 8 February 2019)

Terry, Geraldine (2009) "No climate justice without gender justice," *Gender and Development* **17**(1): 5–18.

Trenberth, Kevin E. (1997) "The use and abuse of climate models," *Nature* **386**: 131–133.

Treuer, Galen, Kenneth Broad and Robert Meyer (2018) "Using simulations to forecase homeowner response to sea level rise in South Florida: Will they stay or will they go?" *Global Environmental Change* **48**: 108–118.

Vitousek, Sean, Patrick L. Barnard, Charles H. Fletcher, Neil Frazer, Li Erikson and Curt D. Storlazzi (2017) "Doubling of coastal flooding frequency within decades due to sea-level rise," *Scientific Reports* **7**(1): 1399.

Wernick, Adam (2017) "Humanity has entered a global warming minefield, climate scientists say," *Public Radio International.* 4 November. Available at: www.pri.org/stories/2017-11-04/humanity-has-entered-global-warming-minefield-climate-scientists-say (accessed: 4 March 2018)

WHO (2017) "Climate change and health". Available at: http://www.who.int/mediacentre/factsheets/fs266/en/

Williamson, Hugh (1770) "An attempt to account for the change of climate, which has been observed in the Middle Colonies in North America," *Transactions of the American Philosophical Society* **1**(1769): 272–280.

Young, Iris (2006) "Katrina: Too much blame, not enough responsibility," *Dissent* **53**(1): 41–46.

Zillman, John and Steven Sherwood (2017) "The enhanced greenhouse effect," *Australian Academy of Science.* Available at www.science.org.au/curious/earth-environment/enhanced-greenhouse-effect

CHAPTER 4

Climate as Determinant

Since man is no independent substance, but is connected with all elements of nature; living by inspiration of the air, and deriving nutriment from the most opposite productions of the earth, in his meats and drinks: consuming fire, while he absorbs light, and contaminates the air he breathes: awake or asleep, in motion or at rest, contributing to the change of the universe; shall not he also be changed by it?

Johann Gottfried Herder, 1800: 164

In his treatise *Outlines of a Philosophy of the History of Man* (1784–1791), the German philosopher Johann Gottfried Herder (1744–1803) offers a conception of the interaction between humans and nature in which environmental and cultural processes are portrayed as distinct yet connected. His words resonate today, at a time when scientists are attending to the embeddedness of society in nature. It is becoming undeniable, as we saw in the last chapter, that human technologies, lifestyles and institutions contribute to "the change of the universe" that Herder refers to (Steffen *et al.*, 2018). At the same time, humans themselves are inevitably affected by the social and environmental conditions they contribute to. Architecture and agriculture, clothing and cuisine, all commonly reflect climatic conditions. But claiming that society is partly affected by the climate is not the same thing as assuming society is wholly determined by it. A common theme of the past has been a less balanced perspective than Herder's, which stressed the dependence of individuals and society on natural conditions, not least on climate. This perspective, known as climate determinism, asserts that human activities are fundamentally and ultimately *controlled* by climate (Livingstone, 2011).

In the decades after the Second World War, climatic determinism was generally dismissed as a naïve view of the world and little intellectual energy was devoted to it. But in recent years the idea has enjoyed something of a renaissance (Hulme, 2011; Stehr and von Storch, 1997). In both its early and later manifestations, climate determinism emphasises the impact of climate on the lives of individuals and social collectives, while restricting the scope of the agency of social institutions, economic inequalities, cultural values and individual choice. In this account, climate works to decide murder rates (Ranson, 2014), migration patterns (Missirian and Schlenker, 2017) and more.

There are some differences, however, between the earlier and contemporary versions of climate determinism: in the early twentieth century ruin and prosperity were both possibilities of the fateful outcome of a controlling climate. The impact of climate was never uniform around the world. Today, in contrast, the emphasis is more often deeply pessimistic and concerned with the cataclysmic widespread consequences of future climate change. The anticipated impact is also usually understood as affecting the entire human species as a whole. Jeffrey Sachs (2012: 258–259), for example, warns "within a generation and probably much sooner [the issues of climate change, water scarcity, resource depletion, and biodiversity] will loom as the largest challenges facing the planet. The world is headed over the cliff, exceeding or soon to exceed the safe global boundaries on countless ecological fronts."

A much more unusual voice is that of Ulrich Beck (2016: 35) who speculates that climate change may well have a significant upside for human civilisation. He calls his view "*emancipatory catastrophism*". See also Beck (2015): "The surprising momentum of metamorphosis is that, if you firmly believe that climate change is a fundamental threat to all of humankind and nature, it might bring a cosmopolitan turn into our contemporary life and the world might be changed for the better". These voices depict climate change as a global future catastrophe facing a homogeneous humanity (see Chapter Five for more on this problematic catastrophic construction).

The possibly dramatic social, economic and political consequences of the forces of climate change already visible and in the not too distant future encourage scientists to probe a rehabilitation of radical climate determinism as a credible perspective on the "nature" of the relation between climate and society. Mike Hulme (2011: 247) warns of "climatic reductionism", which finds expression in some of the bolder claims by scientists and politicians concerning the future impact of anthropogenic climate change. Once more the loss of human agency is seen as the crucial outcome of the impact of climate on society. As Hulme observes (2011: 248): "In this new mood of climate-driven destiny the human hand, as the cause of climate change, has replaced the divine hand of God as being responsible for the collapse of civilisations, for visitations of extreme weather, and for determining the new twenty-first-century wealth of nations."

It is in this context of a possible revival of the idea that climate has a determining factor on human society, that a discussion of the significant tradition of climatic determinism is crucial for an understanding of significant strands of our beliefs and policies on climate and climate change. The hefty appeal and prominence of this tradition is of standalone interest, but it is also pertinent to carefully examine its lacunae and mistakes at a time when the connection between society and "nature" is being re-examined. While this re-examination brings climate back into focus, social values and practices always exceed simplistic reductionist thought. In this chapter we offer a brief survey. We begin by recovering the substance of the ideas, conceptions and pre-occupations of early twentieth century climate determinism. We concentrate particularly on one representative, the geographer Ellsworth Huntington (1876–1947). As we show, early twentieth century climate determinism, as exemplified by Huntington, struggles with various problems. We highlight four in particular: problems of racism, spurious relation, non-falsifiability and lack of agency. In addition, we will review some of the contemporary claims about the impact of climate on civilisation, comparing them with the ideas of classical climate determinism.

While we strongly reject the claims of climate determinists, we also assert that the influence that climate has on society should not be simply disregarded. Thus, we advance the view that it is important to move away from the notion that *climate works* (that is, *immediately controls or decides*) to the idea that *climate matters* in a variety of ways. The perspective of "climate matters" attempts to avoid one of two fallacies in the history of ideas regarding the impact of climate on society (Hulme, 2011). On the one hand the fallacy "of 'climate indeterminism' in which climate is relegated to a footnote in human affairs and stripped of any explanatory power" (Hulme, 2011: 246). On the other hand, the fallacy of climate determinism in which climate becomes everything.

4.1   Climate Works?

For centuries, various scientists, intellectuals, humanists, philosophers and physicians around the globe added weight to the assumption that climate *works*. The notion of "climate works" signals that climatic conditions in particular and environmental context in general are seen to have a heavy causal determining effect on human existence and the rise and fall of civilisations throughout human history. Climate determinism also shaped "attitudes to labour practices, race relations, housing policies, and the management of colonial regimes, sometimes nourishing an imperial mindset, on other occasions underwriting cultural pluralism. It has contributed, often in contradictory ways, to questions about human anatomy and disease, mental health and moral philosophy, medicine and hygiene, plant and human acclimatisation" (Livingstone, 2011: 378).

In the appendix to this chapter we provide a list of suggested climate variables and their alleged effects. The list is most bizarre — its only apparent limitation being the imagination of the thinker. It ranges from conventional measurements such as temperature, humidity and windiness to the exotic such as magnetic storms, concentration of ozone in the atmosphere, number of sunspots or phases of the moon. An enumeration of the effects includes, for example, life expectancy, crime rates, the fall of the Roman Empire, tuberculosis, stock market movements, intelligence, the number of marriages, political revolutions, religious wars, and police arrests.

The impact of climate on humans and societies is believed to have been discussed by the Classical Greek philosophers Aristotle and Hippocrates. Actually, as the research of climate historian Franz Mauelshagen (2018) reveals, the Greek word for "climate" did not appear in their texts and was certainly not a meteorological category in antique geography. Arguably then, the first assertions of climate determinism, and definitely the most famous, appeared in the work of the French philosopher Baron de Montesquieu (1689–1755) who famously announced "l'empire du climat est premier de tous les empires" (see Shackleton (1955)). In his theory of the political division of power (first published in 1748), Montesquieu argued that there is no best form of government but that institutions and justice in a state must harmonise with the given natural conditions and the "nature" of the people. He asserts that observable human variety, for instance, religious doctrines or ethnic diversity is the result of different climatic conditions rather than human biology. Buddhism, for example, is said to be a product of India's climate: Buddha placed his followers in a state of extreme passivity that is reinforced from the laziness of the climate.

For Montesquieu, the influence of climate on the human character became the crucial factor for explaining differing societal and cultural phenomena, such as political institutions, family structures or philosophical systems. According to his theory, people are cognitively and physically more active in cold climate zones than people in warm climate zones. Montesquieu (1748: 248) describes his own observations of climate sensitivity:

> "I have been at the opera in England and in Italy, where I have seen the same pieces and the same performers; and yet the same music produces such different effects on the two nations; one is so cold and phlegmatic, and the other so lively and enraptured, that is seems almost inconceivable."

In this Enlightenment period, discussions about the impact of climate on humans and civilisations were widespread and diverse. One of the Montesquieu's strongest critics was Voltaire (1694–1778). His most meaningful criticism of climatic determinism was that it failed to be able to account for cultural change. For Voltaire the cause of cultural change was linked to moral causes. The German philosopher Georg Wilhelm Friedrich Hegel (1770–1831) pronounced that regions with a torrid or cold climate were not conducive to spiritual freedom, and thus "the temperate zone must furnish the theatre of world history" (Hegel, 1975: 155). The great encyclopaedias of the day considered it a given that ethnic differences were an expression of climatic differences.

By the nineteenth century, the notion that climate played an important, if not determining impact on people and civilisations had been accorded textbook status, as exemplified by the Austrian writer and geographer Friedrich Umlauft (1844–1923) in his work of 1891, *The Atmosphere. Foundations of Meteorology and Climatology According to the Most Recent Research*.

> "And now consider man! ... Because Earth is not merely his living place but also the school of humanity, we must connect racial, national and cultural differences first of all with climate conditions. How different is climate in dealing with people. To some, climate bestows bountiful benefits so that people are enticed into comfortable lightheartedness. Others, forced through the hard school of unavoidable efforts and deprivations, are led to their full development of bodily and spiritual strength ... So the literature of a people is mysteriously connected with the meteorological elements of their part of the globe. The same holds for philosophical teaching systems. So the whole human culture connects with the conditions and processes of the atmosphere. Therefore ... assertions are correct that northern Europe has to thank its rain in all seasons for its position of having the world's highest culture, just as China had in earlier times a high civilisation because of its summer rains." [1]

The career of climate determinism as a major intellectual perspective within the sciences reached its apex, however, in the first two decades of the twentieth century as naturalists, anthropologists, sociologists, physicians and geographers framed a much more quantitative and

---

[1] Our translation.

therefore ostensibly "scientific" approach to the question of the fateful influence of the natural environment on human civilisations and history (cf. Glacken (1967)).[2]

In 1911, the American geographer Ellen Churchill Semple (1911: 1–2), for example, began her widely cited study on the control of human affairs by the natural environment with the following general declaration:

> "Man is a product of the earth's surface ... the earth has mothered him, fed him, set him tasks, directed his thoughts, confronted him with difficulties that have strengthened his body and sharpened his wits, given him his problems of navigation and irrigation, and at the same time whispered hints for their solution ... Man can no more be scientifically studied apart from the ground he tills, or the lands over which he travels, or the seas over which he trades, than polar bear or desert cactus can be understood apart from its habitat." [3]

The doctrine of environmental determinism appeared to offer a solid, broad and scientific foundation that served as the primary explanatory principle regarding the nature of the interaction between environment and people. The assertion that northern Europeans were, in the words of Semple (1911: 620), "Energetic, provident, serious, thoughtful rather than emotional, cautious rather than impulsive," took on an even more pronounced authority that had few, if any, rivals in the social science community or the wider public. The benefits of a "temperate" (as opposed to "tropical" or "polar") climate in underpinning the emergence of civilisation were simply taken for granted.

---

[2] Clarence J. Glacken (1909–1989) was Professor of Geography at the University of California, Berkeley. He is known for *Traces on the Rhodian Shore* (1967), which illustrates how human perceptions of the environment influenced the course of history over millennia.

[3] Semple (1911: vii) speaks of geographic factors and influences and "shuns the word geographic determinant and speaks with extreme caution of geographic control".

This is not to say that all forms of climate determinism were the same. Consider the theory of Semple's teacher, Friedrich Ratzel,[4] who made a distinction between "civilised" races, who occupied the "temperate zones" and "savage" or "natural" races found in the "extreme border regions": "We call them races deficient in civilisation, because internal and external conditions have hindered them from attaining to such permanent developments in the domain of culture as form the mark of the true civilised races." (Ratzel, 1896: 22). His assertions promote the idea that "natural" races were *more* determined by physical environmental environment than "civilised" races, since as "a nation grows it sets itself free from the soil" (Ratzel, 1896: 28). He emphasises that civilised peoples are less dependent upon "individual accidents" of nature and that "natural races" in contrast "are in bondage to Nature" (Ratzel, 1896: 14).

Discussions in this period were influenced by "neo-Lamarckian" conceptions suggesting that when humans were transplanted into a different climate, their physiology actually changed and that the organic consequences of acclimatisation could then be inherited by subsequent generations. The physician and anthropologist Rudolf Virchow (1922: 231), for example, espouses a neo-Lamarckian perspective on climate. Here he is speaking to a gathering of scientists on the problem of acclimatisation:

> "We know that a person who goes out of his fatherland into another country which is markedly different ... perhaps experiences in the first moment a certain animated renewal, but after a short time, mostly after only a few days, feels somewhat uncomfortable, and that he requires a few days, weeks, even months depending on circumstances to find his equilibrium again ... that is something so generally known

---

[4] Friedrich Ratzel (1844–1904) was a German geographer and ethnographer who first employed the term *Lebensraum* ("living space") in the sense that the National Socialists later would. Ellen Semple dedicated her book to Ratzel, writing that she adopted with his permission Ratzel's ideas on environmental determinism for the English-speaking world. Ratzel "performed the great service of placing anthropo-geography on a secure scientific basis" (Semple, 1911:v).

that every man knows and expects it; one assumes that everyone who arrives in such a country and is not completely negligent uses precautionary measures in order to ease this period."[5]

Virchow claims that the human organs literally alter themselves in this phase of acclimatising. This process, as he says, is something like a permanent redressing and may even lead to a climate sickness. The newly formed, climatically adapted organs, according to his conjecture, may become permanent, so that they are bequeathed to descendants. Virchow is writing at a time when colonial expansion is on the political agenda and is concerned that the fertility of individuals who migrate to regions of the world where climates prevail that are different from their "native" climate will suffer. At least in the short run, the population of colonisers is bound to decrease and can only be sustained by a constant influx of new individuals. In the long run, neo-Lamarckians believe that climate can be conquered almost perfectly by way of adaptation and then inheritance.

In contrast, climate determinists who were influenced by Darwinian notions believed that inherited climatic dispositions cannot simply be altered from one generation to another but are at best subject to change in a long-term process of natural selection. This perspective stresses the extent to which climatic conditions attract some individuals while rejecting others. Climatic conditions will assert their superiority and drive out cultural practices that are not in accord with them (cf. Huntington (1945: 610)). In the long run, as Ellsworth Huntington (1927: 165) observes, "Ill health, failure and gradual extinction are the lot of those who cannot or will not adapt themselves to the climate, but before that happens many migrate to other climates better adapted to their physiques, temperaments, occupations, habits, institutions and stage of development."

By the middle of the twentieth century research into climate determinism and its dissemination almost entirely disappeared. In

---

[5] Our translation.

the period after the Second World War the question of the influence of climate on people and society hardly played a role anymore for social scientists. Due to its intellectual and political proximity to racial theories and National Socialist ideology, climate determinism was silenced in the post-War years and largely ignored in natural and social sciences. Climate was assigned the role as a constraint for, rather than as a cause of, human conduct.

However, our observation is that we are seeing a revival of the tendencies of climate determinism today. Reviewing the climate determinism of the early twentieth century is a worthwhile endeavour to understand its flawed premises. For this reason, we turn now to the work of Ellsworth Huntington as perhaps the most prominent advocate of classical climate determinism.

4.2 Ellsworth Huntington

Huntington was a research professor of geography at Yale University and President of the Ecological Society of America in 1917, President of the Association of American Geographers in 1923 and President of the Board of Directors of the American Eugenics Society between 1934 and 1938. There can be little doubt that his ideas regarding climate were connected to his support for, and leadership of, the eugenics movement in the United States.[6] According to Huntington "democracy itself was threatened by the rapid multiplication of the less able members of the species" and he urged "restrictive immigration into the United States" (Martin, 1973: xiv).

The term "eugenics" was coined by the English anthropologist Francis Galton, a cousin of Charles Darwin, who defined it as "the

---

[6] Ellsworth Huntington was not alone among well-known American social scientists who at the time supported the eugenics idea and political movement. The acclaimed economist Irving Fisher who is remembered as the scholar who first devised a working model of the economy was the founder and the first president of the American Eugenics Society. Racial improvements were part of Fisher's economic model (see Aldrich, 1975; Mitchell, 2011: 132–133).

science which deals with all influences that improve the inborn qualities of a race" (Galton, 1904: 1). Its stated aim was to "raise the average quality of our nation" so that "the general tone of domestic, social and political life would be higher" (Galton, 1904: 3). The proponents of this "science" saw it as a route to dealing with the alleged threat of "degeneration" of mankind through emigration and the spread of "feeblemindedness", and they advocated various forms of intervention — breeding, selection, sterilisation, extermination, genocide. Eugenics, as Galton explained, was supposed to "co-operate with the workings of nature by securing that humanity shall be represented by the fittest races" (Galton, 1904: 5). Thus, in his welcoming speech of the Second International Congress of Eugenics in 1921, Henry Fairfield Osborn urged his audience to "delve afresh into nature to restore disordered and shattered society… to know the worst as well as the best in heredity; to preserve and to select the best" (Osborn, 1921: 313). Eugenicists formally proposed "race hygiene" and asserted the right of the state to "safeguard the character and integrity of the race or races on which its future depends" (Osborn, 1921: 312). But different races were hardly regarded as equal. In the grotesque words of Osborn:

"In the United States we are slowly waking to the consciousness that education and environment do not fundamentally alter racial values. We are engaged in a serious struggle to maintain our historic republican institutions through barring the entrance of those who are unfit to share the duties and responsibilities of our well-founded government. The true spirit of American democracy that *all men are born with equal rights and duties* has been confused with the political sophistry that *all men are born with equal character and ability to govern themselves and others,* and with the educational sophistry that education and environment will offset the handicap of heredity. South America is examining into the relative value of the pure Spanish and Portuguese and of various degrees of racial mixture of Indian and Negroid blood in relation to the preservation of their republican institutions." (Osborn, 1921: 312)

Huntington not only took racial categories as biological fact but also had no doubt about the "mental superiority on the part of the white race" (Huntington, 1924a: 31) and that "no amount of training can eradicate the difference" (Huntington, 1924a: 35). But did this mean that one particular race would succeed in any climate? Or were certain races more suited to succeeding in the climate that they had become adapted to? His central concern was how important the climate was as compared to race and other factors in determining the "success" of a civilisation.

In his main treatise, *Civilisation and Climate*, first published in 1915 (the third edition that we reference here was published in 1924), Huntington defends his conviction that climate must be understood as one of the fundamental causative factor, alongside race, in the history of mankind and the rise and fall of civilisations: "Not only at the present, but also in the past, no nation has risen to the highest grade of civilisation except in regions where the climatic stimulus is great." (Huntington, 1924a: 365). When a high level of civilisation appears, he claims, it always coincides with certain climatic qualities (Huntington, 1924a: 12). He explains the appearance of high levels of civilisation where climate today does not possess these qualities, by noting that climate has varied throughout history. He describes variations of climate as "climatic pulsations", which "consist of a shifting of the earth's climatic zones" (Huntington, 1924a: 10). Thus, "So far as climate is concerned, Greece appears to have enjoyed unusually favourable conditions throughout most of the period from 1000 to 300 BC ... and stimulated the Greeks to a high degree of physical and mental energy." (Huntington, 1924a: 22).

Huntington claims that "the direct stimulus of climate" combined with "high racial inheritance due to natural selection" (Huntington, 1924a: 28) underpins the rise of civilisation. But notice that natural selection is also ultimately determined by climate. Climate plays an important part in "selecting certain types of people for destruction or preservation" (Huntington, 1924a: 6). Changes in a climate will "weed out" the less mentally and physically vigorous. This is partly due to the impact of climate upon migration where "a large number of the migrations of history appear to have been more or less directly started by climatic vicissitudes" (Huntington, 1924a: 25). Migration itself is part of the selection process since "the weak, the feeble, the cowardly, and those lacking the spirit of adventure, together with those who lack determination, are gradually weeded out" (Huntington, 1924a: 24). So, although Huntington does not baldly claim that there is one ideal climate, submitting on the contrary that "the optimum climate varies according to a nation's stage of civilisation and also that there is doubtless some difference in the optimum from race to race" (Huntington, 1924a: 17), everything is ultimately determined by climate: "Both by its direct action and through natural selection a warm, monotonous, and unstimulating climate tends to reduce human activity both physical and mental, regardless of race." (Huntington, 1924a: 55).

Huntington's empirical evidence in this book includes statistics of the productivity of factory workers between 1910 and 1913 in Connecticut in the United States. Huntington linked the number of pieces the factory workers produced each month with the average outside temperature. He came up with an ideal outside temperature of about 15°C. His data revealed a slight rise in workers' "efficiency" in January, followed by a constant rise that reaches its maximum in the month of June. In the course of the summer the numbers fall, but around the end of October they reached their peak value again. A similar method was used to analyse the intellectual performances of students and armed forces cadets.

So, what exactly is this "optimal climate"? While temperature is "the most important element" (Huntington, 1924a: 14) *variation* is also crucial in climate, for "people did not work well when the temperature remains constant" (Huntington, 1924a: 15). This variation, however, should be moderate yet "exhilarating". "The failure to appreciate the great importance of the variability in the weather is one of the main reasons why the pervading effect of climate and of changes in the climate is even yet only dimly appreciated." (Huntington, 1927: 142).

Huntington's enumeration of the conditions that make for the best climate for human health, progress and energy therefore is richer than a mere reference to a single factor; he lists a multitude of climatic, seasonal and weather conditions that should be present simultaneously: First, a "fairly strong contrast between summer and winter" with an average temperature of 18°C in the summer and 4°C in the winter. Second, rain in all seasons, "This does not mean constant rain, but enough so that the air is moderately moist much of the time". Third: "constant but not undue variability of weather" (Huntington, 1927: 141).

To illustrate his arguments, Huntington constructed two maps (Figure 4.1) in *The Human Habitat* (1927). The first is a world map of "climatic energy" based on the average temperatures from approximately 1,100 weather stations around the world. The second shows the "civilisation levels" of the regions of the world. These civilisation levels were established on the basis of an inquiry among 50 scholars from 15 countries. Since the two world maps resemble one another, Huntington takes them as "proof" that climate has and will continue to have a decisive influence on the development of civilisation and culture in various regions of the world. As a result of this "deduction", the most stimulating climate in Europe is "proved" to be found in the cities located within a rectangle whose corners are Liverpool, Copenhagen, Berlin and Paris. There is a range of candidates for the best climate on the North American continent, for example, the Pacific Northwest (Seattle, Vancouver), and New Hampshire in New England to New York City; but New Zealand and parts of Australia also have a good climate.

One intriguing set of statistics that Ellsworth Huntington utilises to provide robust empirical "evidence" for his thesis are the circulation figures from 28 public libraries in the United States and Canada.[7] He uses library statistics because, he suggests, they: "[A]fford a vast reservoir of material for a study of intellectual activity." (Huntington, 1945: 345). He groups the libraries into four categories by latitude. In the six most northerly libraries the proportion of all books in circulation that are classified as non-fiction is 55.2 per cent, while the corresponding figure for the eight most southerly cities is 28.9 per cent. What the evidence demonstrates, according to Huntington,

---

[7] More precisely, weighted averages — generally for twenty years, 1920–1939 — of fiction and non-fiction books in circulation.

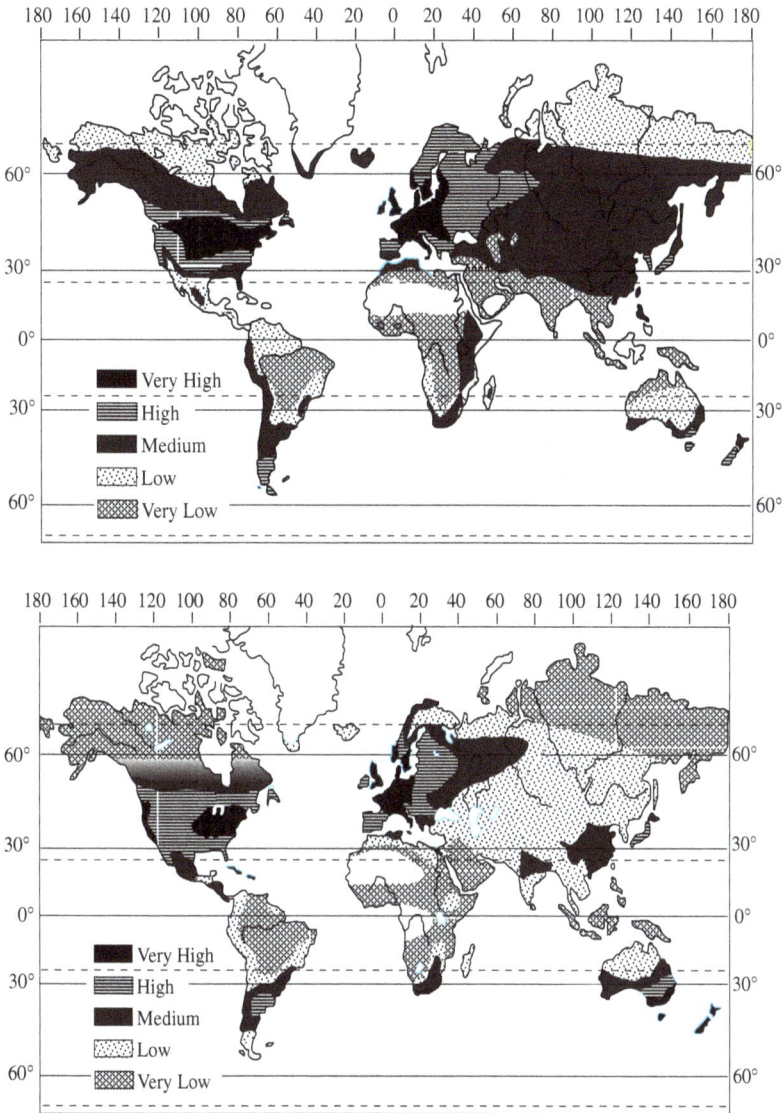

Fig. 4.1. Two key diagrams in Ellsworth Huntington's perspective and analysis. (*Top*) The global distribution of people's energy as derived from the climatic conditions. (*Bottom*) Global distribution of the level of civilisation as derived from a survey among international experts. Both world maps resemble one another — for Huntington, proof that climate has and will continue to have a decisive influence on the civilised and cultural development of various regions of the world.

is that "there is a steady rise from 29 per cent in the most southern cities to 55 in the most northern". But the two intermediate categories also have percentages of non-fiction circulation above 50 per cent (51.3 per cent and 53.5 per cent). This hardly constitutes a "steady rise". Huntington seems oblivious to the fact that a different variable (other than latitude) and a different (and far more convincing) explanation might be relevant to this one anomalous category. Correlation, as all scientists are aware, is not the same as causation.

Huntington's asserted his observations of the determining influence of climate repeatedly. His statements about the importance of climate for the emergence of slavery in the Southern United States, for instance, can be found throughout his various writings covering a period of more than four decades. Whether through repetition, or the influence of those who repeat them, such statements gain credibility and become "common sense" over time. For Huntington, "nothing that man can yet do has any appreciable effect upon the weather, with its changes from day to day and season to season, or upon climate, with its changes in temperature, humidity and wind. On the other hand, *everyone knows* that human feelings, health, and activity are extremely sensitive to weather and climate" (Huntington, 1945: 249, italics added). And "[t]he good sense and scientific temperament of northern people ... are widely recognised" (Huntington, 1945: 344).

Today, Ellsworth Huntington's (1927: 138) tireless work in support of the thesis that "climate paints the fundamental colours on the human canvas" might appear to be extreme or lazy (or even amusing) (Le Roy Ladurie, 1988: 24). However, it was immensely significant at the time for both policymakers and the public and therefore merits careful critique.

4.3 Climate Determinism: A Critique

Various important, detailed and convincing critiques of climate determinism exist (Livingstone, 2002; Meyer, 2000; Hulme, 2011). In this section, we briefly highlight what seem to us to be four specific problematic general features of the texts written by climate determinists in the early twentieth century.

1. First, these theories use climate to construct an "other". It is striking that different authors who agree on the extraordinary influence of climate on human affairs disagree on the specific attributes and the geographic boundaries within which they are supposed to occur and will often advance substantive assertions that are entirely contradictory. Huntington, for example, insists on climate-induced differences between Northerners and Southerners in most countries, whereas Leroy-Beaulieu (1893: 139–144) is convinced that there are discernible convergences in the character of Northern and Southern Europeans because the populations in both regions are subject to climatic extremes. The inconsistency of the claims of these writers brings to light their suspect character. Speculations about the force of climate become an ill-disguised substitute for individual beliefs and prejudice; it is perhaps hardly surprising that "each writer has construed the doctrine so that his own land was regarded as the norm of the temperateness in climate" (House, 1929: 17). British geographer Charles Brooks (from south east England) claims then that "probably South East England has the finest climate" (Brooks, 1951: 30); Leroy-Beaulieu (a European) remarks upon the "great blessing of European nature" (Leroy-Beaulieu, 1893: 8) And Huntington (from the United States) includes parts of the United States alongside England as having conditions approaching "the ideal climate" (Huntington, 1924a: 221). And so on.

   But such theories are not only risible; they are also clearly racist. They have not only favoured a Eurocentric reconstruction of human history, they have also claimed that the future course of world history must necessarily unfold within the limits of its framework, and they construct the racial categories they presuppose. Climate determinism has assisted in developing an overall ideological framework in which racial identity is ascribed to climate and particular races are "naturally" more efficient. Dominant societies are construed as inhabiting favourable climatic regions (with a superior "temperate" climate) while "barbarians" and "uncivilised" people naturally resided in climatically disadvantageous regions (the "tropics" and "polar" regions). Brooks claims that in areas of high temperatures and humidity "the native populations lack energy and initiative and white immigrants cannot maintain their efficiency for many years without recourse to a cooler, more stimulating climate" (Brooks, 1951: 20). Similarly, Robert DeCourcy Ward writes: "In a debilitating and enervating climate ... the will to develop both the man who inhabits the tropics, and also the resources of the tropics, is generally lacking. Voluntary progress toward a higher civilisation is not reasonably to be expected." (DeCourcy Ward, 1908: 227). In his book on Russia, Leroy-Beaulieu explains how extreme cold of that region "disposes to a certain indolence, physical and moral, to a sort of passiveness of mind and soul" (Leroy-Beaulieu, 1893: 141).

These theories work to construct an "other" — the "savage" who is closer to nature, simpler and more inclined to indolence — who is both the product and the victim of their environment. This "other" has low energy and little agency in their own lives and societies. David N. Livingstone notices the persistence of the inclination to construe the climate in *moral* terms and "to use climatic conditions as a vehicle for the transportation of western moral freight" (Livingstone, 2002: 162). He explains the persistence of contrasting the "temperate" world against the "tropical other" (Livingstone, 2002: 160). Climate determinism then has a tendency to construct "the other", who is both simultaneously *differently* determined and *more* determined by the climate and less able to act against it.

2. Another problem is that these hypotheses lack *falsifiability*. There is simply no way to "disprove" the theory of climate determinism. Huntington admits, for example, that "high civilisation" has existed in a different type of climate. But he simply dismisses the apparent refutation of his theory here by arguing that these were merely anomalies: "the effects of climate seem to be overcome for a while, but assert themselves in due time" (Huntington, 1924a: 367). He pores over various maps of the United States and discovers a variety of deviations from the basic pattern of climatic determination but nevertheless concludes that all the maps really show the same basic features and that "the resemblances are too close and too widespread to be accidental" (Huntington, 1945: 24).

Huntington thus always asserts that climate is the only "real" and effectively independent variable in the equation. And where other factors do appear relevant, he argues that they, too, are determined by climate. This both presumes and constructs his basic hypothesis that "social and economic systems everywhere tend to adjust themselves to geographical environment and to the occupations which provide a living in a particular environment at any particular stage of human progress" (Huntington, 1945: 280). He further writes, "Climate operates through soil, vegetation, animals, diet, clothing, housing, disease, and other factors, as well as directly." (Huntington, 1945: 281). A similar claim is made by DeCourcy Ward: "Climate is but one of many controls, albeit a most important one, for it largely determines what many of the other factors, such as diet, customs and occupations, for example, shall be." (DeCourcy Ward, 1908: 223) This means that everything comes back to climate, which works both directly and indirectly. The theory cannot be *disproved*, it cannot be falsified, and for Karl Popper, if a theory cannot be falsified it is unscientific (Popper, 2002: 18).

3. Notice that in these theories, any *correlation* between climate and a social phenomenon is simply and erroneously taken as proving a *causal* connection. Huntington, for example, asserts that there is a causal connection between applications for amendments to patents and seasonal temperature; applications are greatest in the spring and drop in summer (Huntington, 1945: 348–349). The data, he claims "gives an especially reliable picture of seasonal variations in mental activity" (Huntington, 1945: 349). This claim is utterly unconvincing, but even if it were less so, he has not proved that there is a causal connection rather than a "spurious relationship". It may be a coincidence that there is a correlation, or there may be another intervening variable that he has not considered (people tend to go on holiday in the summer, for example). Third, then, climate determinists mistake correlation for causation.

4. Finally, a dangerous and pernicious feature of climate determinism is its reductive eradication of human agency. The implication of these theories is that the self-determined range of human decisions becomes irrelevant, and that history is no longer the result of human subjectivity and activity. Any agency within human decision and behaviour is eradicated, instead they become entirely dependent

on factors that individuals have no influence over whatsoever. People are portrayed as "totally passive, as reacting or submitting to environmental features or events that are thrust upon them" (Franck, 1984: 423). Such an attitude may bolster the status quo; the existing social order must, supposedly for environmental reasons, simply remain as it is. Political authorities are absolved from their role in an environmental crisis or hazard. Undermining this order or violating these rules may be described as threatening or even destroying society's "natural" living conditions and resources.

Underlying climate determinism is a problematic reductive perspective, that is, an assertion of its definitive influence over all attributes and phenomena of a particular situation and a denial that any other aspects of the situation can be of any explanatory importance at all. Here "climate works" to wholly decide the specificities of societies. Thus climate, for determinists, relentlessly imposes its force on humans in an unmediated fashion from which there is no escape: "Not only is our life governed by weather and climate but so is the energy with which we live it." (Brooks, 1951: 11)

Climate is assumed to be responsible for a wide range of human attributes and textures of societies in different regions of the globe. Within each of these climatically controlled societies, it seems there is assumed to be an almost perfect impartiality and equality. The benefits, risks and costs associated with climate and, therefore, the human destinies that are attributed to climate are almost always distributed regardless of social and cultural factors. The uneven distribution of power and capital, understood by sociologists to underpin and result from inequality and stratification, are simply ignored. Climate, it seems, does not discriminate. But this apparently non-selective, unmediated appropriation of climate in mentalities and its direct manifestation in cultural forms and social structures make climate determinism a highly unrealistic description of the interaction between nature and society.

To repeat, we do not deny that environmental conditions such as the availability of natural resources and weather variations affect human conduct, if only as the result of certain social re-constructions of these features in terms of constraints of and opportunities for social conduct, but while they do constitute constraints and opportunities for human conduct, they do not necessarily *determine* it. Opportunities and risks vary historically in their impact; this impact is stratified and as forces that govern social conduct they are sometimes negligible, sometimes critical.

4.4   Climate Determinism: A Revival?

The role of a destabilised climate in destabilising society has had an immense impact within the academic and scientific world and across regions and levels of governance. Scientists, policymakers and the wider public are justifiably concerned with the repercussions of climate change for society. But the sense of urgency has apparently precipitated the beginnings of a remarkable renaissance of climate determinism in contemporary climate research (see Hulme, 2011).

The reappearance of climate determinism marks a shift from the tendencies of post-war science. The social sciences have tended to deliberately discard references to physical, biological and environmental factors. One of the reasons, as noticed by Karen Franck (1984), is that the "ghost" of determinism haunts the social sciences and dissuades them from assigning any influence whatsoever to the physical environment. It may also be because they have shared certain ideological and normative assumptions that the march toward progressive modern societies and desirable living conditions included an extensive emancipation from the immediate impact of and dependence on environmental conditions. The success social scientists have generally enjoyed in dismissing any reference to "natural processes" (except perhaps in the vaguest sense of an insignificant background noise) has been supported for decades by prevalent views in the natural sciences that nature exists in a state of equilibrium and permanence. But in significant parts of the natural science, the concept of "nature" is increasingly losing its static character and "closed-system" attributes; it is described as mutable and dynamic as well as subject to human interference. This means that the interconnection between society and nature is now back at the forefront of many discussions in science and politics, and social science is forced to re-examine its own relation to, and conceptions of, natural phenomena.

Where can such tendencies towards a new "climate determinism" be observed? A vast literature has emerged to assess the impact of increased temperatures on human health (Deschenes, 2014) and agriculture (Schlenker and Roberts, 2009). In this section, however, we focus upon two particular facets of society that have been especially subject to claims of climate determinism — economic activity and socio-political conflict.

First, the specific impacts of a climate changed world are considered in terms of future economic activities. There exists widespread concern that climate change will have a significant impact on economic performance around the world (Diamond, 1997; World Bank, 2008; Sachs, 2001). But there is an important difference between the idea that climate change *matters* and the claim that it is an irresistible influence on society. Some recent analysis, noticeably by economists, veers dangerously close to this claim. Joshua Graff Zivin and Matthew Neidell (2014), for example, argue that weather plays an important role in the marginal productivity of labour. Higher temperatures, they explain, may cause "discomfort, fatigue, and even cognitive impairment" (Zivin and Neidell, 2014: 1) and therefore can lead to changes in time allocated to work especially in climate-exposed industries, such as agriculture, construction and manufacturing. To be sure, they note "our evidence from temperature shocks cannot adequately characterise the behavioural responses that could arise under the more gradual and systemic temperature changes expected under climate change" (Zivin and Neidell, 2014: 23). But the likely plethora of different (non-rational) human responses and the other factors that may influence them are not investigated, nor even mentioned. Their results, as they themselves acknowledge, apply to the short term, and do not take into account structural change over the longer term.

Another example is a recent study investigating the "striking latitude gradient in economic development" or in other words the apparent fact that "as one moves away from the equator in either direction, the level of income per capita goes up" (Andersen *et al.*,

2016). This study argues that ultraviolent radiation (UV-R) — a strong correlate of absolute latitude — also correlates strongly with income and morbidity. The proposed mechanism is that UV-R increases the occurrence of debilitating eye disease, which affects life expectancy and therefore "fertility transition", which is somehow linked to income. The problematic assumptions made in this alleged causative path are not probed and nor are the other, possibility more relevant social factors, such as the legacy of colonialism and uneven global power relations.

Second, there is a growing area of research investigating the relations between climate and political conflicts. Hsiang and Burke, for example, "find consistent support for a causal association between climatological changes and various conflict outcomes, at spatial scales ranging from individual buildings to the entire globe and at temporal scales ranging from an anomalous hour to an anomalous millennium (Hsiang and Burke, 2014: 39). See also Hsiang *et al.* (2011) and Burke *et al* (2009). Burke and colleagues conclude: "Anthropogenic climate change has the potential to substantially increase global violent crime, civil conflict and political instability, relative to a world without climate change." (Burke *et al.*, 2015: 610). The onset of the Syrian conflict, for example, is frequently said to be linked to a multi-year drought in the region (see Selby, 2014).

A historical analysis is offered by Murat Iyigun and colleagues (2017), who retrospectively analyse the effect of cooling on conflicts in Europe, North Africa and the Middle East between 1400 and 1900. The authors claim to be able to show that cooling is associated with an increase in conflict, especially in the long run. There is ample historical evidence, the authors suggest, that "the reduction in agricultural productivity that resulted from cooling led to different types of conflict. There are examples of peasant rebellions in times of famine" (Iyigun *et al.*, 2017: 11). Some, however, actually propose the opposite relation: Salehyan and Hendrix (2014: 239), for example, using water scarcity as their empirical case, argue that there are "good reasons why water scarcity might have a pacifying effect on armed conflict, and that political violence should be more prevalent during periods of comparatively better agro-climatic conditions". Between 1970 and 2006, an abundance of water rather than adverse environmental conditions is correlated with political conflict. The authors claim that political conflicts are *more* likely to happen when basis existential needs are met.

However, noticeably, the exact underlying mechanism for these apparent correlations remains unknown (Hsiang and Burke, 2014) and the methodological reflections are currently limited (cf. Ide (2017)). Moreover, this research suffers from a "streetlight effect" — "if the evidence of a causal association between climate and violent conflict is informed only by exceptional instances where violent conflict arises and climate also varies in some way, it is unable to explain the vastly more ubiquitous and continuing condition of peace under a changing climate (in neighbouring societies)" (Adams *et al.*, 2018).

Other critics have pointed to the problematic temporal and spatial assumptions, and vastly oversimplify both political conflict and the phenomena of climate change (Selby, 2014). In the "climate-conflict problematic", the role of socio-political institutions, power relations and economic exploitation is overlooked (Selby, 2014: 19). Indeed, one analysis suggests that "[o]ur results suggest an urgent need to reform African governments' and foreign aid donors' policies to deal with rising temperatures" (Burke *et al.*, 2009). Here, other institutions and the bigger picture are omitted from the picture. Next, we consider this twenty-first century version of climate determinism in comparison to the earlier version.

4.5   Climate Determinism: A Comparison

As Hegel and Marx may remind us, a socio-historical development is rarely a radical break with the past in which old tendencies vanish completely and an entirely new ones emerge as if from nowhere (Dahrendorf, 1959: 29). We might say that climate determinism in its twenty-first century manifestation *transcends* classical climate determinism; some elements of previous versions have been abolished, and some preserved. It is notable, however, that references to classical climate determinism are usually absent from contemporary studies.[8]

Of course, the social, political, economic and scientific context of contemporary climate determinism is distinct from that within which classical climate determinism flourished and found a strong resonance; a context of colonialism and a different scientific understanding of climate and climate change. Today, the background is rather one of growing alarm at both the hugeness of the potential impact of climate change and the paucity of the general response to this potential impact, particularly in the global north.

As we examine in depth in the Chapter Three, the exact impact of a changing climate remains uncertain and contested, but anthropogenic climate change seems likely to have a significant, if not uneven impact upon living conditions around the globe — both in the future and in the present. Extreme weather events such as heat waves, droughts and forest fires are said to be "the new normal"; although what is actually meant by "normal" here is that there *is* no longer any "normal".[9] Scientists refer to what we are experiencing as a "Hothouse Earth" (Steffen *et al.*, 2018). On a global scale, the years 2016, 2017, 2015 and 2014 were the four warmest years ever measured worldwide, in that order (see NOAA, 2018). Climate change, it is therefore argued, in line with climate determinism, will have a significant impact on societies. The question is how *mediated* this impact is by human individuals and social institutions. Take one recent statement: "Although climate is clearly not the only factor that affects social and economic outcomes, new quantitative measurements reveal that it is a major factor, often with first-order consequences." (Carleton and Hsiang, 2016).

As we go on to examine more closely in Chapter Six, many in the scientific community and beyond are dismayed by the lack

---

[8] For example, Clarence Glacken's *Traces on the Rhodian Shore* (1967) with its extensive examination of nature and culture in Western thought is without any reference to the ideas of his fellow geographer Ellsworth Huntington.

[9] https://www.cbsnews.com/news/are-devastating-wildfires-a-new-normal-its-actually-worse-than-that-climate-scientist-says/

of efforts to respond to these worrying trends (see Rich, 2018). Publics and governments seem either unwilling or unable to put in place satisfactory policies (see Marshall (2014)). Even the 2015 Paris Accord, widely regarded as a historical achievement and as marking a general consensus on the threat of anthropogenic climate change, is non-binding and unenforceable. The Treaty's goal is to limit warming to two degrees; the odds that this will be achieved based on current emissions has been *estimated* as one in twenty (Raftery *et al.*, 2017). NASA scientist James Hansen argues that all the Paris Accord marks is an agreement that there is a problem, not an agreement on what to do about it. He states in a recent interview with *The Guardian* that "the world is failing 'miserably' with the worsening dangers".[10] There is an urgency here that stems from possible long-term consequences. And yet the institutions of liberal democratic governance seem to be easily captivated by the immediacy of frequently and rapidly changing "events", and this is facilitated by constitutional rules of representation that prescribe relatively short frames of temporality.

Could it be that the sense of urgency and concomitant frustration has led to a tendency to impress the drastic and determining nature of the climate change? Perhaps versions of modern climate determinism, articulated precisely to highlight the inadequacy of a "business as usual" approach, are perceived as justified for a good cause? But it might be that the motivation for the new climate determinism also stems from incentives internal to the scientific community. The search for establishing the basis for anthropogenic climate change has largely been accomplished. Could it be that efforts to study the range of changes that may be attributable to climate change have become the new frontier of climate researchers?

Contemporary narratives of climate determinism do not seek to differentiate people on the basis of climate, rather they lump together a humanity that is simultaneously both victim and villain of the

---

[10] See https://www.theguardian.com/environment/2018/jun/19/james-hansen-nasa-scientist-climate-change-warning

"civilisation-shaking catastrophe".[11] "Not one of us is innocent, not one of us is safe," writes Ray Scranton (2018). He urges a radical change towards less environmentally destructive ways of life. Many accounts position humanity as a species suffering together as havoc wreaks the planet and who must act together to save it. Social inequalities and power relations do not play a big part in this depiction. So, whereas accounts such as Huntington's use *past* climate to determine present social distinctions, these contemporary narratives use *future* climate to obliterate them; humanity is framed against the background of the cataclysmic climate change of tomorrow and everything else is left out of the picture.

And yet despite these differences, these new versions of climate deterministic nonetheless suffer from the four problematic tendencies highlighted earlier. Arguably, contemporary climate determinism seems just as reductive and inattentive to human agency, just as difficult to falsify, and just as guilty of confusing correlation with causation, as its precursor. There is also the issue of the way in which such determinism constructs the "other". While these theories certainly do not use racial categories, they do portray certain populations as being helpless refugees or violent culprits with little choice about their role in social disorder, since we are told, "altered environmental conditions stress the human psyche, sometimes leading to aggressive behaviour" (Hsiang *et al.*, 2011). Selby argues that there is a tendency to reproduce stereotypes of impoverished victims and perpetrators of conflict living in the global South (Selby, 2014: 19).

Any simple extrapolation from past to the future and from climate (change) to society must be called into question as neglecting complex pathways and interrelationships. As William Meyer writes: "How climatic change matters will always depend upon how society has evolved and continues to evolve, just as will the significance of those aspects of climate that remain the same." (Meyer, 2000: 213).

---

[11] See https://www.theguardian.com/environment/2018/jun/19/james-hansen-nasa-scientist-climate-change-warning

4.6   Conclusion: Climate Matters

Contemporary social science has for the most part successfully avoided the seductive simplicity of most forms of environmental and biological determinism. The history of the social sciences in the last century can be written as a struggle against various forms of determinism, and mainstream social science has succeeded in restricting its focus to social, political, economic or cultural processes. The ecosystem, refashioned to a lesser or greater extent by social action (particularly by way of appropriating its resources, for example), remains a major material condition and constraint for human conduct. But pointing to the relevance of environmental aspects in the complex dynamics of society is not the same thing as arguing that climate decides its fate. The notion of nature in social science discourse must be reconsidered in a manner that avoids the pitfalls of any reductionist climate determinism.

Due to the revelations from both natural and social sciences, nature is moving from a background concern into centre stage; increasingly, "climate matters". But it is not enough to merely introduce the *topic* of environment into social science discourse. The discipline of environmental sociology, for example, has been behind initial as well as more sustained efforts in recent years to re-introduce environmental conditions into social science discourse (Grundmann *et al.*, 2012). But for the most part this has been a plea to incorporate ecological topics into social theory and thus thereby recognise that society affects the environment. The environment here is constituted as a problem that can be dealt with as an external consideration, through existing social, political and economic institutions. The environment continues to be located *outside* society; environmental problems do not disrupt the internal structure of society. Moreover, society itself plays no role in the construction of these problems.

But there exist, however, many voices from various disciplines attesting to the idea that climate and society are not external to each other; the "environment" should not be perceived as only a limitation or condition of society, but as itself conditioned by society. As Noel Castree and Bruce Braun write in their preface of their book

Social Nature (2001), geographers, anthropologists, cultural analysts, ethnographers and philosophers have all contributed to "the idea that nature is social" and overturned the assumption that the natural and the social were two distinct domains. For not only is what we understand as "natural" *physically* transformed by society, as proponents of the Anthropocene diagnosis reveal, but the very category "nature" is also itself *ideologically* constructed by society. What is understood as "natural" is contingent upon a cultural and social context. This is not to deny the very real existence of icebergs and thunderstorms, but it is to claim that classifying them as "natural" depends on social discourse. Are the new species that are emerging as a result of climate change entirely natural, or in someway "social" too? Can we position *ourselves*, as members of the human species, as outside nature? As Anna Peterson (2001) points out, the notion that human beings are somehow discontinuous with the natural world commonly features in Western cultures, is not a universal position. Many alternative understandings to "human exceptionalism" exist outside the West. Conceptions of the connection between "nature" and "human" are themselves highly significance for ethical systems and cultural practices (Peterson, 2001).

Emphasis upon this interconnectedness reminds us that "nature", however conceived, sculpts the society that contemplates and constructs it: "While humans build the environments in which we live, work, worship and play, those environments, in turn, shape our understandings of ourselves." (Gabrielson and Parady, 2010: 381). In other words, we should be careful in our turn away from climate determinism not to assume the opposing position. "Social constructionist" accounts too simplistically reverse the determinism, by reducing climate to culture. Here, nature is seen as itself a cultural formation; for social constructionists "the one thing that is not 'natural' is nature herself" (Soper, 1995: 7). As Kate Soper stresses, it is possible to resist the appeal of both naturalism and constructivism; it is also possible to deny that "nature" is an entirely objective force and to avoid reducing it entirely to ideology.

We propose a shift away from the presupposition of "climate works" towards a notion of "climate matters". This shift involves a firm refusal to succumb to the seductive simplicity of climate determinism and its construction of "otherness", its lack of falsifiability, its confusion between correlation and causation and its pernicious denial of human agency. But it also acknowledges the interconnections between climate and society. Climate clearly affects our social reality and our cultural horizons.

We see the interconnection of the natural climate with social reality, particularly in the ways in which society responds to climate extremes. Climate extremes are institutionalised and inscribed into society, for instance, in the form of a wide variety of myths, ideologies, stories, technologies, regulations and organisations. An obvious and powerful example is protective levees built at rivers and oceans as well as the laws and regulations that govern their construction, maintenance and use. In much the same way, the evolution of shelter and clothing are to some extent an inscription of climate into the social fabric. These artefacts constitute responses to climate and are, to a certain extent, "portraits" of social encounters with climate. But these efforts and technologies impact both social interactions with the climate system.

Take modern means of transportation, for example. Utilised to link open spaces with each other and to carry commodities, information and human beings, transportation takes place in familiar and tight enclosures that keep out undesirable climatic conditions. Nonetheless, the climatic conditions from which we desire to withdraw are engraved into these enclosures. As time and distance become increasingly irrelevant in a globalising world, the influence of extremes on the construction of such artefacts also increases. Paradoxically, these extremes tend to vanish from view; climate becomes nearly invisible. But such means of climate-indifferent transportation also emit the greenhouse gases that contribute to climate change. Climate thus demonstrates its "reality" in the social conduct that comprehends and conditions it. Climate matters to the societies partially situated within it.

As Wilhelm Lauer puts it (1981: 24) "Climate shapes the theatre in which human existence … takes place, sets borders for that which can happen on the earth, but certainly does not determine what happens or will happen. Climate introduces problems that man (sic) has to solve. Whether he solves them, or how he solves them, is left to his imagination, his will, and his formative activities." Our point, however, is that these formative activities impact deeply engrained understandings of society, humanity and culture and that the solutions designed to tackle climate problems may reverberate across the social-nature nexus in unforeseen ways. Climate matters, for sure, but how precisely it matters is a topic for further research.

Appendix

The Efficacy of Climate — Past and Present. An Inventory

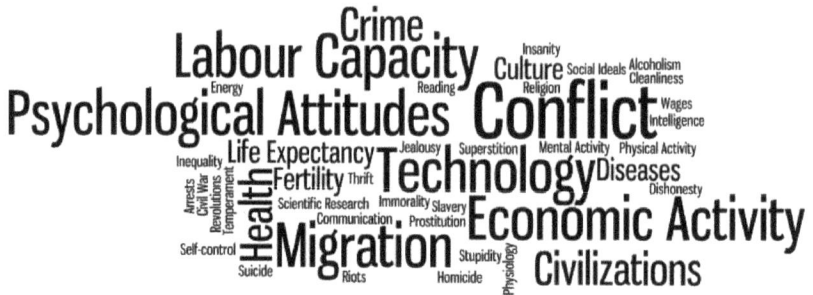

Crime
Labour Capacity Culture Social Ideals Alcoholism
Insanity Cleanliness
Energy Reading Religion
Psychological Attitudes Conflict Wages
Intelligence
Inequality Life Expectancy Jealousy Superstition Mental Activity Physical Activity
Fertility Thrift Technology Diseases
Dishonesty
Scientific Research Immorality Slavery
Communication Prostitution Economic Activity
Self-control Migration Stupidity Civilizations
Suicide Riots Homicide Physiology

This inventory collates many of the alleged causal linkages between climate and social phenomena past and present.

Alcoholism (Semple, 1911: 626)

Arrests (Huntington, 1945: 363–364)

Civilisations ("As the Tropics have been the cradle of humanity, the Temperate Zone has been the cradle and school of civilisation. Here Nature has given much by withholding much." (Semple, 1911: 635); Figure 86 'Map of Civilisation' on page 256 in Huntington and Cushing (1921: 256); "The distribution of civilisation throughout the world has always depended closely upon climate." (Huntington, 1927: 165); "By encouraging one type of social organisation and discouraging another, climate has great influence upon the development of civilisation." (Huntington, 1945: 276); "The greatest events of universal history and especially the greatest historical developments belong to the North Temperate Zone." (Semple, 1911: 611); "Where man has remained in the Tropics, with few exceptions he has suffered arrested development." (Semple, 1911: 635); "The question has been repeatedly raised as to whether there have been changes in climate in historical times, especially rainfall fluctuations, sufficient to explain the decline and fall of the Roman Empire and the decadence of civilisation, by reason of which large sections of the Mediterranean lands, once thriving and populous, have become depopulated or impoverished. Chiefly historians, archaeologists, and other incompetent authorities not concerned with climatology have advanced arguments supporting this position. The majority of competent authorities have reached a contrary conclusion ... Ellsworth Huntington attributed the decline of Palestine, Syria, Asia Minor, Greece and Italy to the same cause, but his arguments have been questioned both by historians and climatologists." (Semple, 1931: 99–100); "A map of climate, or rather of climatic energy, as we may call it, resembles a map of progress far more closely than does a map of any other factor which may be a cause rather than a result of the distribution of progress." (Huntington, 1927: 140); see also Hsiang and Burke (2014: 46f) on the decline and collapse of ancient Chinese dynasties.)

Cleanliness ("The climate itself may also be largely responsible for the lack of cleanliness [in this case, among Icelanders]. So far as I am aware, this lack prevails among every people who live in a cool, moist climate where the water is always cold and where animals are the chief means of support ... The cleanest people in the world are the inhabitants of warm, moist countries, where the state of culture requires clothing, and where there is plenty of water." (Huntington, 1924a: 289))

Communication (as dependent on favourable climatic conditions, e.g., Huntington (1924a: 300))

Conflict ("Studies best positioned to make rigorous causal claims overwhelmingly indicate strong linkages between climatic anomalies and conflict and social instability. Furthermore, this growing literature demonstrates that climate's influence on security persists in both historical and modern periods, is generalisable to populations around the globe, arises from climatic events that are both rapid and gradual, and influences numerous types of conflict that range across all spatial scales." (Hsiang and Burke, 2013: 52; Hsiang *et al.*, 2011: 438; Burke *et al.*, 2015; Iyigun *et al.*, 2017; Beine and Parsons, 2017))

- **Civil war** (in the United States: "In all these respects climatic contrasts paved the way for civil war." (Huntington, 1945: 280; Burke *et. al.*, 2009; Iyigun *et al.*, 2017))

- **Revolutions** ("In the world as a whole the tendency toward lack of self-control in politics, in sex relations, and in many other respects rises markedly in hot weather and in hot countries. This is not the only reason for the frequency of political revolutions in low latitudes, but it must play a part." (Huntington, 1945: 365; Iyigun *et al.*, 2017))
- **Riots** ("Weather as a promoter of riots has hitherto been neglected. Nonetheless it seems to agree with the distribution of riots [in India],"; "it is noteworthy that in the United States Negro riots occur most often in unusually hot weather." (Huntington, 1945: 362, 364) ("Bohlken and Sergenti (2010) and Sarsons (2011) find that negative rainfall growth is associated with higher risk of Hindu-Muslim riots in Indian states over the last few decades of the twentieth century." (Hsiang and Burke, 2014: 50; Burke *et al.*, 2015))

Crime (Huntington, 1945: 365–367); for a modern version of the same claim, see Ranson (2014) who claims that between 2010 and 2090, *climate change* in the United States will cause an additional 22,000 murders, 180,000 rapes and 1.2 million aggravated assaults. See Burke *et al.* (2009) and Cohen and Gonzalez (2018).

- **Sexual crimes** (Huntington, 1945: 365)
- **Homicide** ("Homicide shows a significant relation to temperature both geographically and seasonally... seasonally as well as geographically, the rates increase from cooler to warmer weather ... warm weather apparently is associated with lowered self-control. It also makes people feel disinclined toward steady effort. Lack of self-control is a primary factor in promoting murder. Disinclination to work is a primary factor in the failure of public sentiment to express itself in observance of law." (Huntington, 1945: 232); see also Burke *et al.* (2015))

Culture ("Climate ... helps to influence the rate and the limit of cultural development. It determines in part the local supply of raw material with which man has to work, and hence the majority of his secondary activities, except where these are expended on mineral resources. It decides the character of his food, clothing, and dwelling, and ultimately of his civilisation." (Semple, 1911: 609); "The North Temperate

Zone is preëminantly the cultural zone of the earth." (Semple, 1911: 634); "Cultural variations from season to season seem to be intimately connected with physiological conditions that manifest themselves in reproduction and in rate of work." (Huntington, 1945: 319)).

Diseases (The impact of climate on health is stressed by many climate determinists, even though it may only be in a kind of superficial and less consequential fashion — see below — than the stronger assertion that infectious diseases of one sort or the other are either promoted or repressed by climatic conditions: "Climate undoubtedly modifies many physiological processes in individuals and peoples, affects their immunity from certain classes of diseases and their susceptibility to other." (Semple, 1911: 608; Andersen *et al.*, 2016))

Economic activity

- **Commerce** (the decline and rise of commercial activities as dependent on climate, e.g., Huntington (1924a: 300)).
- **Economic cycles** ("The rhythm in the activity of economic life, the alternation of buoyant, purposeful expansion with aimless depression, is caused by the rhythm of the yield per acre of the crops; while the rhythm in the production of the crops is, in turn, cause by the cyclical changes in the amount of rainfall. The law of the cycles of rainfall is the law of the cycles of crops and the law of economic cycles." (Moore, 1914); "Almost every advanced country has sharp seasonal variations in its occupations, wages, trades, transportation, bank clearings, and other phases of business." (Huntington, 1945: 312); "Prices depend on many condition … it also confirms the idea that a full understanding of fluctuations in business is impossible unless we understand how much is due to environmental cycles in contrast to purely economic or human reactions." (Huntington, 1945: 488)).
- **Economic preferences** Becker *et al.* (2018) examine the global variation of economic preferences, in particular their link "to the structure of mankind's ancient migration out of Africa"; the authors attempt to document that "these temporally distant migration movements have shaped today's heterogeneity in risk, time, and social preferences, both across and within countries, albeit to

heterogeneous degrees across preferences"; and the results "establish that the effect of temporal distance on differences in risk aversion and prosocial traits is robust to an extensive set of covariates, including controls for differences in the countries' demographic composition, their geographic position, geographic distance metrics, **prevailing climatic** [the proxies are temperature and precipitation] and agricultural conditions, institutions, and economic development." (Becker *et al.*, 2018: 1, 3; emphasis added).

- **Economic prosperity and development** ("Economic prosperity and general well-being are distributed according to much the same geographical pattern as social welfare." (Huntington, 1945: 232); "One of the sturdiest regularities in comparative economic development involves the location of a country vis-à-vis the equator and its level of prosperity. As one moves away from the equator in either direction, the level of income per capita goes up." (Andersen *et al.*, 2016: 1334, 1361; Burke *et al.*, 2015: 611; Iyigun *et al.*, 2017; Beine and Parsons, 2017))
- **Efficiency** ("Extremes both of heat and cold reduce the density of population, the scale and efficiency of economic enterprises." (Semple, 1911: 611))
- **Wages** ("The low cost of living keep down [the] wages, so that the labourer ... is poorly paid [in southern countries and regions] ... The labourer of the north, owing to his providence and larger profits, which render small economies possible, is constantly recruited into the class of capitalist." (Semple, 1911: 620–621; Beine and Parsons, 2017)).

Energy (Figure 85 on page 255 in Huntington and Cushing (1921), "Map of Climatic Energy", shows how "human energy would be distributed if it depended wholly on climate"; the map sums up the "combined effects of temperature, humidity, seasons and storms upon health and energy" (Huntington, 1927: 145); "The energy and progress of the world's leading countries is due to the constant repetition of the physiological stimulus which comes with the changing seasons." (Huntington, 1945: 319)).

Fertility (Virchow, 1922: 231); The reproductive "cycle varies according to climate", "In the northern United States and western Europe the maximum of births normally occurs in March or April as a response to conceptions in June or July. Elsewhere the maximum tends to shift to earlier dates in hot climates and later ones where the climate is cold." (Huntington, 1945: 273–274; Andersen *et al.*, 2016)

Health: One of the more frequently cited effects of the climate is that on health. ("The climate of Iceland is not only healthful but stimulating." (Huntington, 1924a: 289); "The geographical distribution of health and vigour depends largely on the combined effect of climate and cultural conditions." (Huntington, 1945: 240); "In the United States infants conceived in the fall and born in the summer are especially numerous and have the lowest percentage of congenital defect." (Huntington, 1945: 319); "The resistance of infants ... to digestive diseases apparently varies according to their age in a way that suggests an innate adaptation to a particular kind of climate. The peculiar ability of people, especially women, in the reproductive ages of life to resist disease during the late winter suggests the same thing." (Huntington, 1945: 610; Graff *et al.*, 2014: 24; Obradovich *et al.*, 2017))

Inequality ("The old South distinguished sharply between aristocrats and 'poor whites', as well as between whites in general and Negroes. This distinction of classes was in strong contrast to the relative democracy that prevailed in the North, where the squire might care for his own horse, cow and garden. When slavery disappears, a system of tenancy almost invariably grows up in regions where differences in ability to manage people and property are especially important in comparison with the ability to do manual work." (Huntington, 1945: 367; also Andersen *et al.* 2016)).

Insanity ("At that time [June] the physical stimulus which merely leads to health and increased powers of reproduction among normal people apparently overestimates those who are poorly poised, weak of will, oversexed, or otherwise abnormal." (Huntington, 1945: 365))

Intelligence ("People of high latitudes are, on the whole, more intellectual than those of low latitudes." (Huntington, 1945: 367))

Labour capacity ("Differences in health indicate corresponding differences in inclination to work, as well as in actual capacity to work.

Vigorous people prefer to work rather than sit idle. The will to work beyond the required limits is extremely important in crisis, such as war, flood, or other disaster. It is one of the main factors in leading people to make inventions, explore new lands, carry out scientific experiments, initiate reforms, and produce works of art, literature, and music." (Huntington, 1945: 238); "A hot climate, especially if it is humid, makes people feel disinclined to work. This encourages the more clever [sic.] people to get a living with as little physical exertion as possible. Their example fosters the growth of a social system in which hard work is regarded as plebeian." (Huntington, 1945: 276); "The greatest social influence [of climate] is probably its effect on inclination to work." (Huntington, 1945: 282; Graff *et al.*, 2014: 2; Andersen *et al.*, 2016; Burke *et al.*, 2009))

Life expectancy "June as we have seen is a good time of good health and maximum conceptions, especially in Western Europe. Children conceived then or a little earlier live longer and are more likely to be eminent than those conceived at any other time." (Huntington, 1945: 365; 1945: 610); ("Bodily temperatures rises [in the Torrid Zone], while susceptibility to disease and rate of mortality show an increase ominous for white colonisation." (Semple, 1911: 626; Iyigun *et al.*, 2017))

Migration ("The acclimatisation of tropical people in temperate regions will never be an equation of widespread importance ... [The Negroes'] concentration in the 'black belt', where they find the heat and moisture in which they thrive, and their climatically conditioned exclusion from the more northern states are matters of local significance. Economic and social retardation have kept the hot belt relatively underpopulated." (Semple, 1911: 625–626); "the people in poorer climates are practically certain to have poorer health and less energy than others. The population as a whole is likely to be less prosperous, so that education and contact with other people are less prevalent. Moreover, under such circumstances there is a strong tendency for the more able people to leave the poorer environment." (Huntington, 1927: 162); "Climatic conditions begin to mould and select the migrants to the new environment." (Huntington, 1927: 165; 1945: 184). For a modern version, linking *climate change* to migration, see Reuveny (2007), Graff *et al.*

(2014: 25); Iyigun *et al.* (2017), Beine and Parsons (2017), Cattaneo and Bosetti (2017) and Missirian and Schlenker (2017).

Physical activity ("Physical vigour is basic in human progress ... Vigour is needed in order that people may work hard without undue fatigue and have a reserve of strength in emergencies. It is especially import-ant in promoting mental activity and clear thinking." (Huntington, 1945: 237); "Physical vigour is one of the main factors in the growth of civilisation." (Huntington, 1945: 275); the "optimum temperature depends upon the conditions under which man took the evolutionary steps which gave him his present adjustment to climate" (Huntington, 1945: 273); "At temperatures above the optimum, fatigue is readily induced, the inclination to work diminishes, and the easiest way to make oneself conformable is to do as little as possible. At temperatures below the optimum the inclination to work is stimulated, partly because bodily activities promote warmth, partly because there are many ways in which a moderate degree of inventiveness enables people to keep themselves warm artificially." (Huntington, 1945: 275). See Graff *et al.* (2014) for an account of the changes in outdoor leisure activity due to climate)

Physiology ("The effects of a tropical climate are due to the intense heat, to its long duration without the respite conferred by a bracing winter season, and its combination with the high degree of humidity prevailing over most of the Torrid Zone. These are conditions that are advanta-geous to plant life, but hardly favourable to human development. They produce certain derangement in the physiological functions of heart, liver, kidneys and organs of reproduction." (Semple, 1911: 626)).

Prostitution "seem to reach a maximum in the hottest parts of the world, that is, in the dry parts of a belt located ten to thirty degrees from the equator" (Huntington, 1945: 296).[12]

---

[12] Huntington (1945: 296) refers, in this context to Hellpach (without further specificity; however, in the bibliography, Willy Hellpach's 1911 *Die geopsy-chischen Erscheinungen des Wetters, Klima und Landschaft in ihrem Einfluss auf das Seelenleben* is listed) and quotes him as saying that "in Southern Italy sexual irregularities increase greatly when the sirocco is blowing. The peo-ple recognise this so well that offenses committed under such circumstances are in a measure condoned."

Psychological Attitudes ("People feel growingly optimistic in the spring and still more so in the autumn." (Huntington, 1945: 318))

- **Dishonesty** ("The climate of many countries seems to be one of the great reasons why idleness, dishonesty, immorality, stupidity, and weakness of will still prevail." (Huntington, 1924a: 411))
- **Jealousy** ("The men of the hot desert may have unusual cause for jealousy. In extremely hot weather people's ability to resist emotional impulses, including those of sex, appears to be weakened. Sexual extravagance and prostitution seem to reach a maximum in the hottest parts of the world, that is, in the dry parts of a belt locate ten to thirty degrees from the equator." (Huntington, 1945: 296))
- **Mental activity** ("Among European races physical activity appears to be the greatest when the temperature averages not far from 65° F, whereas mental activity seems to be greatest at a lower temperature, averaging perhaps 40°." (Huntington, 1924a: 290); in addition, climate variability stimulates mental activity, e.g., Huntington (1924a: 290))
- **Self-control** (Climatic "extremes weaken the power of self-control" (Huntington, 1924a: 404); there is "evidence that dry weather, especially when hot, is associated with a decline in self-control". (Huntington, 1945: 296))
- **Temperament** ("The northern peoples of Europe are energetic, provident, serious, thoughtful rather than emotional, cautious rather than impulsive. The southerners of the sub-tropical Mediterranean basin are easy-going, improvident except under pressing necessity, gay, emotional, imaginative, all qualities which among Negroes of the equatorial belt degenerate into grave racial faults." (Semple, 1911: 620); "The important point is that people's temperament fluctuates in harmony with the weather ... There is a widespread impression that hasty legislation and personal violence in the form of fist fights formerly rose to a maximum [in the Congress of the United State before air conditioning was introduced] under such conditions. It is worth noting that in the United States Negro riots occur most often in unusually hot weather." (Huntington, 1945: 364)). ("Furthermore, altered environmental conditions stress the

human psyche, sometimes leading to aggressive behaviour." (Hsiang *et al.*, 2011: 440))

Reading (Huntington, 1945: 391)

Religion ("Diversity of physical environment has also been effective in leading to religious differences, and among the environmental factors climate has been especially important." (Huntington, 1945: 281); "Protestantism appears to be the phase of Christianity best adapted to regions in which the physical as well as cultural conditions are highly energising, so that people do a great deal of thinking and feel inclined to make strenuous efforts, both physical and mental." (Huntington, 1945; 302)).[13]

Slavery ("It was not only the enervating heat and moisture of the Southern States, but also the large extent of their fertile area which necessitated slave labour, introduced the plantation system, and resulted in the whole aristocratic organisation of society of the South." (Semple, 1911: 622); "Slavery failed to flourish in the North not because of any moral objection to it, for the most godly Puritans held slaves, but because the climate made it unprofitable." (Huntington, 1924a: 41); "The suppression of slavery in the North was not due chiefly to moral conviction. That arose after long experience had shown that slavery did not pay in a cool climate ... the combination of good food, stimulating climate, and northern type of culture made the white

---

[13] Since religious belief systems are not merely other-worldly but from this world, early mythological and later more systematic religious belief systems always display certain environmental constraints with which their origina-tors struggled and they even tend to reflect or incorporate certain climatic conditions (Hoheisel, 1993) but this is of course a far cry from maintain-ing an almost indiscriminate assertion that religious beliefs and practice are driven by climatic conditions. Moreover, as Hoheisel (1993: 130) points out, available ethnographical information lack reliability and validity to clearly tie religious beliefs and practices to climatic conditions: "In any case, in-creasing spatial mobility and progressive liberation from natural constraints, for example through trading and commerce at a distance, but above through the possibility of being able to build on traditions of very different origins, make it considerably more difficult to prove that beliefs or other religious doctrines are characterised by certain climatic conditions."

northerners so energetic that it irked them to wait for slow-moving Africans." (Huntington, 1945: 279))

Scientific research ("... the world's scientific research and other intellectual activities, as well as its financial, commercial, industrial and political control are more and more becoming concentrated in the few limited regions where the climate is most healthful and stimulating." (Huntington, 1927: 160))

Social norms and ideals ("The difference in inclination toward work had much to do with the development of diverse social ideals in these parts of the United States. In the North the successful family was the one where everybody worked hard as well as intelligently. Hard work became the supreme virtue, as it is to this day in spite of other tendencies. In the South the successful ante-bellum family was one that eschewed physical labour and at the same time got a good living. This system favoured slavery and attached a social stigma to work with the hands. An aristocratic society was almost inevitable, because the mental ability to get a good living through slave labour is more limited than the physical ability which was so important in the North." (Huntington, 1945: 280))

Suicide ("In 1922 four California cities led the list of suicides ... Possibly these facts may be connected with the constant stimulation of the favourable temperature and the lack of relaxation through the variations from season to season and from day to day, although other factors must also play a part. The people of California may perhaps be likened to horses which are urged to the limit to that some of them become unduly tired and break down." (Huntington, 1924a: 225; 1945: 365))

Superstition: (e.g., Huntington, 1924b: 297)

Technology

- **Inventions** (Huntington, 1945: 391; Iyigun et al., 2017)
- **Patent productivity** ("An isoplethic [or 'contour'] map I have made, of American patent productivity per capita, shows a heavy

concentration in the narrow belt of best climate, near the 50°F isotherm, from Chicago to Philadelphia and Boston." (Gilfillan, 1970: 46)).

- **Patent applications or rejections** (Kovács, 2017) "This paper explores how weather variations influence patent examination decisions at the USPTO. It demonstrates that higher than normal temperature corresponds to an increase in allowance rates and a decrease in final rejection rates, while higher than normal cloud coverage corresponds to lowered final rejection rates. These weather effects are partly due to a sorting effect, whereby on warmer days examiners tend to work on higher quality patents. Even after controlling for this sorting effect, however, weather variations influence allowance and final rejection rates in the above described ways." (Kovács, 2017: 1833).

Thrift ("The necessity of preparing shelter, clothing and fuel as means of combating the cold and moisture of winter tends to promote a social system which places high value on foresight and thrift." (Huntington, 1945: 277))

References

Adams, Courtland, Tobias Ide, Jon Barnett and Adrien Detges (2018) "Sampling bias in climate–conflict research," *Nature Climate Change*.

Aldrich, Mark (1975) "Capital theory and racism. From laissaz-faire to the eugenics movement in the career of Irving Fisher," *Review of Radical Political Economics* **7**: 33–42.

Andersen, Thomas Barnbeck, Carl Johan Dalgaard and Pablo Selaya (2016) "Climate and the emergence of global income differences," *Review of Economic Studies* **83**: 1335–1364.

Beck, Ulrich (2016) *The Metamorphosis of the World*. Cambridge: Polity.

Beck, Ulrich (2015) "Emancipatory catastrophism: What does it mean to climate change and risk society?" *Current Sociology* **63**: 75–88.

Beine, Michel and Parsons, Christopher R. (2017) "Climatic factors as determinants of international migration: Redux," *CESifo Economic Studies* **63**(4): 386–402.

Brooks, Charles (1951) *Climate in Everyday Life*. New York: Philosophical Library.

Burke, Marshall, Solomon M. Hsiang and Edward Miguel (2015) "Climate and conflict," *Annual Review of Economics* **7**: 577–617.

Burke Marshall, Edward Miguel, Shanker Satyanath, John A. Dykema and David B. Lobell (2009) "Warming increases the risk of civil war in Africa," *PNAS* **106**: 20670–206774.

Carleton, Tamma A. and Solomon M. Hsiang (2016) "Social and economic impacts of climate," *Science* **353**(6304): 9837.

Castree, Noel and Bruce Braun (2001) *Social Nature: Theory, Practice and Politics*. Malden, Oxford, Victoria: Blackwell.

Cattaneo, Cristina and Bosetti, Valentina (2017) "Climate-induced international migration and conflicts," *CESifo Economic Studies* **63**(4): 500–528.

Cohen, François and Gonzalez, Fidel (2018) "Understanding interpersonal violence: The impact of temperatures in Mexico," *Grantham Research Institute on Clime and the Environment*. Working Paper No. 291. Available at: http://www.lse.ac.uk/GranthamInstitute/publication/understanding-interpersonal-violence-impact-temperatures-mexico/. (Accessed 22 March 2018)

Dahrendorf, Ralf (1959) [1957, 1st ed.] *Class and Class Conflict in Industrial Society*. Stanford, California: Stanford University Press.

DeCourcy Ward, Robert (1908) *Climate: Considered Especially in Relation to Man*. New York: G. P. Putnam's Sons.

Deschenes, Olivier (2014) "Temperature, human health, and adaptation: A review of the empirical literature," *Energy Economics* **46**: 606–619.

Diamond, Jared (1997) *Guns, Germs, and Steel. The Fates of Human Societies*. New York: Norton.

Franck, Karen A. (1984) "Exorcising the ghost of physical determinism," *Environment and Behaviour* **16**(4): 411–435.

Gabrielson, Teena and Katelyn Parady (2010) "Green citizenship: rethinking green citizenship through the body," *Environmental Politics*, **19**(3): 374–391.

Galton, Francis (1904) "Eugenics: Its definition, scope, and aims," *American Journal of Sociology* **10**(1): 1–25.

Gilfillan, S. C. (1970) [1935, 1st ed.] *The Sociology of Invention: An Essay in the Social Causes, Ways and Effects of Technic Invention, especially as Demonstrated Historically in the Author's Inventing the Ship*. Cambridge: MIT Press.

Glacken, Clarence. J. (1967) *Traces on the Rhodian Shore: Nature and Culture in Western Thought from Ancient Times to the End of the Eighteenth Century*. Berkeley: University of California Press.

Glacken, Clarence J. (1992) "Reflections on the history of Western attitudes to nature," *GeoJournal* **26**: 103–111.

Grundmann, Reiner, Markus Rhomberg and Nico Stehr (2012), "Nature, climate change and the culture of the social sciences". In: *Rethinking Climate Change Research*. (Eds.) Pernille Almlund, Pernille, Per Homann Jespersen and Søren Riis. Clean-Technology, Culture and Communication. London: Ashgate, pp. 133–142.

Hegel, Georg Wilhelm (1975) [1830, 1ˢᵗ ed.] *Lectures on the Philosophy of World History*. (Trans.) H. B. Nisbet. Cambridge.

Herder, Johann Gottfried von (1800) [1784–1791, multivolume,1ˢᵗ ed.] *Outlines of a Philosophy of the History of Man*. Translated from the German *Ideen Zur Philosophie der Geschichten der Menschheit* by T. Churchill. New York: Bergmann Publishers.

Hoheisel, Karl (1993) "Gottesbild and Klimazonen". In: *Studium Generale 1992*. Heidelberg: Heidelberger Verlagsanstalt, pp. 127–140.

House, Floyd N. (1929) *The Range of Social Theory. A Survey of the Development, Literature, Tendencies and Fundamental Problems of the Social Sciences*. New York: Henry Holt.

Hsiang, Solomon M. and Marshall Burke (2014) "Climate, conflict, and social stability: What does the evidence say?" *Climatic Change* **123**: 39–55.

Hsiang, Solomon M., Kyle C. Meng and Mark A. Cane (2011) "Civil conflicts are associated with the global climate," *Nature* **476**: 438–441.

Hulme, Mike (2011) "Reducing the future to climate: A story of climate determinism and reductionism," *Osiris* **26**: 245–266.

Huntington, Ellsworth (1945) *Mainsprings of Civilisation*. New York: John Wiley and Sons.

Huntington, Ellsworth (1927) *The Human Habitat*. New York: Van Nostrand.

Huntington, Ellsworth (1924a) [1915, 1ˢᵗ ed.] *Civilisation and Climate*. 3ʳᵈ ed. New Haven: Yale University Press.

Huntington, Ellsworth (1924b) *The Character of Races as Influenced by Physical Environment, Natural Selection and Historical Development*. New York, London: C. Scribner's Sons.

Huntington, Ellsworth and Sumner W. Cushing (1921) *Principles of Human Geography*. New York: John Wiley & Sons.

Ide, Tobias (2017) "Research methods for exploring the links between climate change and conflict," *WIREs Climate Change* 8, 1–14.

Iyigun, Murat, Nathan Nunn and Nancy Qian (2017) "Winter is coming: The long-run effects of climate change on conflict, 1400–1900," NBER Working Paper 23033. Available at: http://www.nber.org/papers/w23033

Kovács, Balázs (2017) "Too hot to reject: The effect of weather variations in the patent examination process at the United States Patent and Trademark Office," *Research Policy* **46**: 1824–1835.

Lauer, Wilhelm (1981) *Klimawandel und Menschheitsgeschichte auf dem mexikanischen Hochland* Abhandlungen der Mathematisch-Naturwissenschaftlichen Klasse, Nr. 2. Mainz, Wiesbaden: Akademie der Wissenschaft und der Literatur, pp. 1–50.

Le Roy Ladurie, Emmanuel (1988) [1967, 1st ed.] *Times of Feast, Times of Famine: A History of Climate Since the Year 1000*. New York: Farrar, Strauss and Giroux.

Leroy-Beaulieu, Anatole (1893) *Empire of the Tsars and the Russians*. Volume 1. New York: Putnam.

Livingstone, David N. (2011) "Environmental determinism". In: *The Sage Handbook of Geographical Knowledge*. (Ed.) John A. Agnew. Sage: Los Angeles, pp. 368–380.

Livingstone, David N. (2002) "Race, space and moral climatology: Notes toward a genealogy," *Journal of Historical Geography* **28**: 159–180.

Marshall, George (2014) *Don't Even Think About It. Why Our Brains Are Wired To Ignore Climate Change*. New York: Bloomsbury.

Martin, Geoffrey J. (1973) *Ellsworth Huntington. His Life and Thought.* Hamden, Connecticut: The Shoe String Press.

Mauelshagen, Franz (2018) "Climate as a scientific paradigm — Early history of climatology to 1800". In: *The Palgrave Handbook of Climate History.* (Eds.) Sam White, Pfister, Christian and Mauelshagen, Franz. Basingstoke: Palgrave Macmillan.

Meyer, William B. (2000) *Americans and their Weather.* New York: Oxford University Press.

Missirian, Anouch and Wolfram Schlenker (2017) "Asylum applications respond to temperature fluctuations," *Science* **358**: 1610–1614.

Mitchell, Timothy (2011) *Carbon democracy: Political power in the age of oil*. Verso Books.

Montesquieu, Baron de (1748/2001) *The Spirit of Laws*. Ontario: Batoche Books.

Moore, Henry L. (1914) *Economic Cycles. Their Law and Cause*. New York: Macmillan.

NOAA (2018) "Global climate report June 2018," Available at: https://www.ncdc.noaa.gov/sotc/global/201806. (Accessed 9 August)

Obradovich, Nick, Robyn Migliorini, Sara C. Mednick and James H. Fowler (2017) "Night time temperature and human sleep loss in changing climate,"

Science Advances **3**(5). Available at: http://advances.sciencemag.org/content/3/5/e1601555/tab-article-info. (Accessed 22 March 2018)

Osborn, Henry Fairfield (1921) "The second international congress of eugenics address of welcome," *Science* **54**(1379): 311–313.

Peterson, Anna L. (2001) *Being Human. Ethics, Environment, and Our Place in the World*. Berkeley, California: University of California Press.

Popper, Karl (1959) *The Logic of Scientific Discovery*. London, New York: Routledge, 2002.

Raftery, Adrian E., Alec Zimmer, Dargan M. W. Frierson, Richard Startz and Perian Liu (2017) "Less than 2°C warming by 2100 unlikely," *Nature Climate Change* **7**: 637–645.

Ranson, Matthew (2014) "Crime, weather, and climate change," *Journal of Environmental Economics and Management* **67**: 274–302.

Ratzel, Friedrich (1896) *The History of Mankind*. (Trans.) A. J. Butler. New York: Macmillan and Co.

Reuveny, Rafael (2007) "Climate change-induced migration and violent conflict," *Political Geography* **26**: 656–673.

Rich, Nathanial (2018) "Losing earth: The decade we almost stopped climate change," *New York Times*, August 1. Available at: https://www.nytimes.com/interactive/2018/08/01/magazine/climate-change-losing-earth.html?rref=collection%2Fsectioncollection%2Fscience&action=click&contentCollection=science&region=rank&module=package&version=highlights&contentPlacement=7&pgtype=sectionfront. (Accessed 9 August 2018)

Sachs, Jeffrey (2012) *The Price of Civilisation. Reawakening Virtue and Prosperity After the Economic Fall*. London: Vintage Books.

Sachs, Jeffrey (2001) "Tropical underdevelopment," *NBER Working Paper* 8119, DOI: 10.3386/w8119.

Salehyan, Idean and Cullen S. Hendrix (2014) "Climate shocks and political violence," *Global Environmental Change* **28**: 239–250.

Schlenker, Wolfram and Michael J. Roberts (2009) "Nonlinear temperature effects indicate severe damages to US crop yields under climate change," *Proceedings of the National Academy of Sciences* **106**: 15594–15598.

Scranton, Roy (2018) *We are Doomed. Now What? Essays on War and Climate Change*. New York: Soho.

Selby, Jan (2014) "Positivist climate conflict research: A critique," *Geopolitics* **19**: 829–856.

Semple, Ellen Churchill (1931) *The Geography of the Mediterranean Region. Its Relation to Ancient History*. New York: Henry Holt and Company.

Semple, Ellen Churchill (1911) *Influences of Geographic Environment, on the Basis of Ratzel's System of Anthropo-geography*. New York: Holt, Rinehart and Winston.

Shackleton, Robert (1955) "The evolution of Montesquieu's theory of climate," *Revue Internationale de Philosophie* **9**(33/34): 317–329.

Soper, Kate (1995) *What is Nature?* Oxford: Wiley-Blackwell.

Steffen, Will, *et al.* (2018) "Trajectories of the earth system in the Anthropocene," *Proceedings of the National Academy of Sciences of the United States of America.*

Stehr, Nico and Hans von Storch (1997) "Rückkehr des Klimadeterminismus?" *Merkur* **51**: 560–562.

Umlauft, Friedrich (1891) *Das Luftmeer. Die Grundzüge der Meteorologie und Klimatologie nach den neuesten Forschungen gemeinfasslich dargestellt.* Leipzig: Hartleben's Verlag.

Virchow, Rudolf (1992) [1885, 1st ed.] "Über Akklimatisation". In: *Rudolf Virchow und die deutschen Naturforscherversammlungen*. (Ed.) Karl Sudhoff. Leipzig: Akademische Verlagsanstalt, pp. 214–239.

World Bank (2008) *World Development Report 2009. Reshaping Economic Geography*. Washington, DC: World Bank.

Zivin, Joshua and Matthew Neidell (2014) "Temperature and the allocation of time: Implications for climate change," *Journal of Labor Economics* **32**: 1–26.

CHAPTER 5

Climate as Public Perception

… [T]o us weather is linked to unstableness or variability, climate to stableness, changing from location to location only, but not from one time to another. People's awareness of the invariability of the climate is deep-rooted and is expressed in the firm conviction that unusual weather conditions in one season or year will have to be compensated for in the next.

Eduard Brückner, 2000: 78

More than a century ago, Eduard Brückner wrote about anthropogenic climate change and its impact upon society (Stehr *et al.*, 1995). One notable feature of his analysis was his arguably still pertinent observation that a common understanding of climate is based on a strong belief in its *stability*. Brückner points to the dominance of an apparently deep-rooted conviction that climate is a set of conditions consistently following the diurnal and annual cycles that provide the seasonal differences in every part of the world.

Does Brückner's claim still hold? It seems that despite various warnings from scientists (including Brückner) over the centuries, of the variability and unpredictability of the climate system, climate remains widely understood to be a set of constant conditions that remains stable. Climate commonly appears as a reliable resource, a background against which life can be planned in advance and take place as planned. This conviction does not exclude the possibility, of course, that some degree of disruption is inevitable; unusual dramatic events such floods, heat waves and hurricanes take place, but the expectation is that conditions will return to "normal" sooner or later, so that habitual routines can eventually resume. The disruption to normality is itself normal.

In this book, we have focused mostly on the *scientific* understanding of climate so far. In this chapter we attend instead to the way that climate and climate change are understood more generally in society. We notice that our form of life is based upon, and reproduces, the assumption of a constant climate in which we place our trust. Trust functions to allow social interaction to take place. As Niklas Luhmann notices: "We put our trust in the self-evident matter-of-fact 'nature' of the world" — and in fact this propensity to trust *itself* seems to constitute a natural feature of the world (Luhmann, 1979: 4). By trusting that climate will remain constant, society can bracket off unknown futures. For Luhmann, trust is needed to counter unpredictability and thus it "risks defining the future" (Luhmann, 1979: 20); trust is, thus, ultimately a risky practice. Indeed, paradoxically, trust in the constant conditions of climate may serve to veil the ways in which societies are contributing to the destabilisation of those very conditions.

The message from climate scientists today is that the stability of the climate can no longer be taken for granted. Whereas once the everyday trust in climate coincided with the natural scientific construction of climate as a stable global object, today's science reveals the potent volatility of climate, such that prevailing scientific opinions contradict many everyday assumptions. The risks and challenges of climate change have moved to the forefront not only of scientific research but also of much media analysis and political discourse. And although the reality of anthropogenic climate change is frequently and passionately refuted, its assertion can nevertheless hardly be avoided. As a feature of growing concerns about global warming, another narrative of climate has become dominant, that of climate as a source of catastrophe. According to the proponents of this catastrophic depiction, global warming looms on the horizon as a future crisis that will wreak havoc across the entire planet. In this narrative, the climate should be resolutely distrusted.

In this chapter we attend to these two prominent depictions: climate as *constant* and climate as *catastrophe*. We ponder whether these two apparently distinct depictions actually both function to encourage the same social response, or rather a *lack* of response, to the issue of climate change. We turn then to the role of climate scientists in the policymaking process and consider the issue of trust and mistrust of science. We argue that perceptions of climate and climate change are deeply entwined with most forms of life, and although they are never rigidly fixed, they nevertheless cannot be easily displaced. It is unlikely that they will be budged by either invoking illusions of cataclysmic futures on the one hand, or scientific truths on the other.

5.1 Trusting the Climate

A widely held belief in the scientific community during much of the twentieth century was that climate was stable. As Stefan Brönnimann explains this was connected to the rise of "technological futurism": "Climate control appeared feasible, and even if climate cannot be fully controlled, then humankind would easily master the consequences." (Brönnimann, 2015: 259). See also Stehr and von Storch (2000: 12).

This construction of climate as *constant* is something that helps us to plan our daily lives and future activities; we assume that temperatures, precipitation and the occurrence of storms and heat waves will follow the pattern of the year before. As George Perkins Marsh writes in an early environmentally conscious text:

> "Nature left undisturbed, so fashions her territory as to give it almost unchanging permanence of form, outline, and proportion, except when shattered by geologic convulsions; and in these comparatively rare cases of derangement, she sets herself at once to repair the superficial damage, and to restore, as nearly as practicable, the former aspect of her dominion." (Marsh, 1864: 27).

A constant climate, one which follows an equilibrium, has become a firm taken for granted feature of our forms of life. In this section we consider the way in which trust in the stability of climate allows social institutions and interaction to take place and the way in which this trust may ultimately become dysfunctional (Stehr and Machin, 2016).

Trust has been understood to be crucial for allowing social institutions to exist. Various scholars have noted that trust is needed for social interactions, political institutions and economic exchanges to function (Barber, 1983; Fukuyama, 1995; Offe, 1999; O'Neill, 2002; Putnam, 1995; Sen, 1999). "Society operates on some basic presumption of trust." (Sen, 1999: 39) As Claus Offe explains, trust is a resource that allows social coordination to operate. "Trust is the cognitive premise with which individual or collective/corporate actors enter into interaction with other actors." (Offe, 1999: 45). Trust relations exist horizontally between individuals, and vertically between citizens and elites. Economic exchange depends upon trust, which removes the need for formal contracts and reduces what economics call "transaction costs". (Sen, 1999: 263–265; Fukuyama, 1995). Trust has been understood to be a form of social capital that facilitates social cooperation and civic participation (Putnam, 1995).

Trust is not only placed in individuals but in social institutions. For Anthony Giddens trust is *a continuous state* (Giddens, 1990: 32) connected to the feeling of *ontological security*. He is referring to the confidence that humans have in "the constancy of the surrounding social and material environments of action" (Giddens, 1990: 92). Without trust we would be in a state of constant angst, preoccupied with the "low-probability high-consequence risks" (Giddens, 1990: 133) that for Giddens characterise modernity, and which would otherwise prevent us from getting on with our daily lives, rendering us unable even to get up in the morning (Luhmann, 1979: 4). Trust reduces the complexity of everyday life. It limits the range of possible choices and generates a taken-for-granted background against which reflection can take place and decisions can be made. As Kenneth Newton (2007: 351–352) observes: "Every time we visit the dentist, the doctor, the bank, a restaurant, a shop, or a cinema, and every time we use a lift, travel on a bus, train, or plane, even cross a bridge or walk in the street, we trust the individuals, equipment, infrastructure, and public regulation of these things."

Crucially, trust allows us to plan ahead. "To show trust..." writes Luhmann, "... is to anticipate the future. It is to behave as though the future were certain." (Luhmann, 1979: 10). Only if one can fully anticipate future conditions is trust not required as the basis for practical action, but of course such foresight is impossible. Trust allows us to tolerate uncertainty (Luhmann, 1979: 15). Trust thus involves, as Georg Simmel stresses, a combination of knowledge and ignorance: "Confidence, as the hypothesis of future conduct, which is sure enough to become the basis of practical action, is, as hypothesis, a mediate condition between knowing and not knowing another person. The possession of full knowledge does away with the need of trusting, while complete absence of knowledge makes trust evidently impossible." (Simmel, 1906: 450). Or as Giddens puts it more succinctly: "Trust is only demanded where there is ignorance ... yet ignorance always provides grounds for scepticism or at least caution." (Giddens, 1990: 89).

But it is important to notice that trust is not necessarily a rational undertaking, where we have carefully considered the contingencies with which we are faced. Trust is more than uncalculated simple hope, but it "can be shown to be thoughtless, careless and routinised, and this require no unnecessary expenditure of consciousness ... Trust merges gradually into expectations of continuity, which are formed as the firm guidelines by which to conduct our everyday lives". (Luhmann, 1979: 24–25). This means that trust might well be a gamble, "a risky undertaking". (Luhmann, 1979: 26)

Trust, we suggest, is not only an essential ingredient of social, economic and political life that involves our fellow human beings and basic institutions, but also extends to features of our *natural* environment.[1] Trust in stable "natural" conditions, provides a foundation for security and enlarges options for social action. In the face of the complexity and uncertainty of the social world, trust that the climate will remain stable, may provide us with a measure of systemic security and certainty. It seems that the trust that is extended toward climatic and other environmental processes provides an important bracketing and therefore offers room for action and reflection with respect to other pressing issues in everyday life.

---

[1] This is what Hume called the "uniformity of nature" assumption (see Okasha (2016)).

Trust in the stability of the climate becomes significant when the expectations generated from trust make a difference in the decision making of individuals and collectives. Trust about the continuation of the present into the future excludes alternatives for social action and political decision. In short, trust is crucial, but at the same time, it can become dysfunctional when it limits the articulation and visibility of perspectives that might constitute important new possibilities. Trust in stable climate conditions against which we plan our lives might actually serve to block an awareness that the way in which humans live their lives may be undermining such conditions.

Today, however, is everyday trust in the stability of climate not coming under increasingly pressure? Does (as we consider later in the chapter) the accumulation of scientific research or the widespread media images of starving polar bears not serve to disrupt the "common sense" idea of the stability of climate? Might the concrete experience of warmer temperatures and unusual or extreme weather provoke greater public recognition of a more capricious climate? If, as Matt Shardlow remarks, nature appears to "skip a season" (Shardlow, 2016) then will we not start to *distrust* the construction of climate as a constant condition? Research has shown, on the contrary, an apparently widespread apathy regarding climate change amongst the public, at least in parts of the global north (Giddens, 2009; Egan and Mullin, 2012; Myers *et al.*, 2021; Norgaard, 2011; Moser and Dilling, 2007).

In a recent paper on the topic of the public perception of climate change, James Hansen and colleagues ponder if, and how, everyday assumptions of climate will change in response to the real changes in weather patterns in parts of the world. Their main argument is that the obstacle of public recognition of climate change "is probably the natural variability of local climate". They ask: "How can a person discern long-term climate change, given the notorious variability of local weather and climate from day to day and year to year?" Their answer: "The climate dice is now loaded to a degree that a perceptive person old enough to remember the climate of 1951–1980 *should*

recognise the existence of climate change, especially in summer."
(Hansen *et al.*, 2012: E2415, emphasis added). But the presupposition
here is that conceptions regarding climate simply originate in concrete
past experiences and that present deviations should disrupt these
understandings. This is of course a highly speculative assertion about
personal experience with climate and climate change, which not only
assumes that people have a reliable memory of past weather patterns
and that new experiences deviate from such patterns, but that this will
also directly alter common convictions regarding the climate and unpin
our trust in its continuity.

What it is key to recognise here, is that an experience with *weather*
does not translate directly into knowledge of the *climate*. As we
discussed in Chapter Two, weather and climate are distinct phenomena.
On the contrary, it is precisely this *lack* of correlation that reveals their
distinctness, as well as the robustness of the construction of climate
as a set of constant conditions. Climate and climate change cannot be
directly experienced. While it might be likely that an extreme weather
event is provoked by a destabilised climate system, this cannot be
finally proven. In Chapter Three we explained that weather always
varies, for different reasons, and that the notion of a "normal" climate is
a scientific construction. It is worth considering whether, paradoxically,
framing an event as a "weather extreme" can actually provide renewed
confidence in a stable and "normal" climate. Might the labelling of
an event as "*unusual* weather" extreme serve to provide reassurance
that extremes are events that digress from the *usual* climate, therefore
reaffirming that there is such a thing? The exception is precisely what
proves the rule. Trust in constant climate is momentarily destabilised
but ultimately reassured.

Expectations and perceptions of climate are shaped by social and
cultural contexts; many forms of life are based upon a trust upon the
stability of climate and not its volatility. What then might disrupt this
trust? In the next sections we consider the role of catastrophic narratives
and scientific data in constructing and disrupting social perceptions
of climate and climate change.

5.2 Distrusting the Climate?

Over the last few decades, climate change has become a prominent issue; as we saw in Chapter Three, scientists no longer regard the climate as a set of static conditions but as a volatile system replete with tipping points and feedback loops that make future conditions difficult to predict. Perhaps as a result of both the serious changes measured and of the desire to draw attention to them, climate change is often presented as a planetary crisis.

In apparent contrast to the widespread everyday understanding of climate as *constant condition*, is a newly dominant depiction of climate as a source of future *catastrophe* (see Machin, 2013: 116–123). Images of melting ice caps, rising sea levels, flooding, pollution, pestilence, war and general misery appear throughout the media and popular culture. Recent headlines read: "2020 deadline to avert climate catastrophe" (Hood, 2017) and "Climate Catastrophe, Coming Even Sooner?" (Kolbert, 2016). This type of imagery is distributed not only by journalists and filmmakers, it is expressed by scientists and politicians too. Earth scientist James Lovelock, for example, claims: "The consequences for humanity could be truly horrific, if we fail to act swiftly the full impact of global warming could cull us along with vast populations of the plants and animals with whom we share Earth…. there might — in the twenty-second century — be only a remnant of humanity eking out a diminished existence in the polar regions and the few remaining oases left on a hot and arid Earth." (Lovelock, 2011). The "climate crisis" is similarly described by paleontologist Tim Flannery as the: "[o]ne problem facing humanity […] so urgent that, unless it is resolved in the next two decades, […] will destroy our global civilisation." (Flannery, 2009: 14). And US vice president Al Gore writes: "Our climate crisis may at times appear to be happening slowly, but in fact it has become a true planetary emergency and we must recognise that we are facing a crisis." (Gore, 2006: back cover).

Susanne Moser and Lisa Dilling observe the prevalence of "hard-hitting fear appeals" designed to "evoke apocalyptic fears in lay audiences … Words, images, tone of voice, and background music convey danger, darkness, pending doom." (Moser and Dilling, 2007:

164). They note that these sorts of appeals are not just articulated in political campaigning, but in serious scientific discussion too.

It is hoped, presumably, that such dystopian visions will provoke society into action before it is too late. Imagery of imminent crisis is expected to disrupt our taken-for-granted trust in the stability of climate. When he writes that "we stand today on a precipice of annihilation" for example, Roy Scranton (2018) is trying to provoke radical change. But while these types of statements might well convey the potentially devastating impact of climate change and the urgency to address it, do these depictions of climate as a source of catastrophe really provoke a robust social response? Does it engage political actors and ignite collective action? Or might the invocation of catastrophe serve to do precisely the opposite?

Such depictions might actually induce a feeling of helplessness, powerlessness and guilt, which help to generate a fatalistic attitude towards the changing climate. As environmental sociologist Kari Norgaard observes, in the global north the negative emotions around the issue, that are uncomfortable partly because they violate social norms, can motivate people to keep the issue "at a distance". The strange result is that: "the possibility of climate change is both deeply disturbing and almost completely submerged, simultaneously unimaginable and common knowledge". (Norgaard, 2011: xix).

Brigitte Nerlich and Rusi Jaspal are also concerned with the negative emotional responses to climate change. They examine the visual images that often accompany news stories about extreme weather events, such as floods, droughts, hurricanes and heat waves. They explain that most of these images had "emotional meanings of fear, helplessness and vulnerability and, in some cases, guilt and compassion" which they see as *negative* passive emotions as opposed to "active emotions linked to engagement and responsibility". As they point out, rather than provoking positive behaviour change, images that symbolise threat are more likely to lead, on the contrary, to "denial and paralysis" (Nerlich and Jaspal, 2014: 272). Moser and Dilling write: "The principal problem with fear as the main message of climate change communication, is that what grabs attention (dire predictions, extreme consequences) is often not what empowers action." (Moser and Dilling, 2007: 164).

Notice that a consistent feature of the descriptions quoted above is the articulation of *humanity as a whole*, facing the chaos and collapse that looms on the horizon. All humans are victims, regardless of their different geographical, economic and social positions. The "apocalyptic events" invoked by Lovelock (2011: vii) do not seek out some more than others, although — as we have seen in the Chapter Three — the impact of a changing climate will be unevenly distributed. And "we" are not only the victims, but also the villains; humanity as a whole is also construed as entirely responsible for this fate of a struggling planet. "We have chosen the worst of times to add to its difficulties." (Lovelock, 2011: viii). In order to "safeguard and strengthen the social, environmental and economic fabric of life on our shared planet," states one report, "we must now act with great urgency to fulfil our shared commitment … failure is simply not an option." (Revill and Harris, 2017: 7).

Not only is this imperative rather overwhelming, it seems to preclude any real discussion regarding alternative pathways and options, and the various ramifications of climate policy that will inevitably not suit everyone. The danger is that rather than provoking an engagement with a complex, multi-dimensional problem profoundly interconnected with society and its forms of life, the imagery of climate change as a catastrophe closes down options and seems to beckon a solution that will "save humankind from itself". For Eric Swyngedouw, such climate catastrophism problematically counterposes a homogenous humanity against a hostile nature, in which any conflicts of interest between different sectors of society are disavowed (Swyngedouw, 2011: 221): "The presentation of climate change as a global humanitarian cause produces a thoroughly depoliticised imaginary, one that does not revolve around choosing one trajectory rather than another, one that is not articulated with specific political programs or socio-ecological project or revolutions." (Swyngedouw, 2011: 219). He explains that instead of inviting different possibilities that might draw together social movements and political collectives bent on socio-economic change, the spectre of climate change as catastrophic implies that nothing really has to alter: "We have to change radically, but within the contours of the existing state of the situation." (Swyngedouw, 2011: 219).

It may be, then, that the use of totalising catastrophic imagery can actually undermine its intended purpose to provoke political engagement and innovative policymaking with the issue of climate change. Perhaps the combined effect of extreme weather events, the reaching of "tipping points" (such as a massive release of methane brought about by the thawing of the permafrost regions of the world), the accumulation of scientific data and the persistent messages of climate activists may significantly alter public perceptions of the urgency and the risks that may come with a volatile climate. There may be a growing demand for political representatives to take action. But it may result in precisely the opposite affect: the instigation of paralysis and the lack of any political engagement whatsoever.

5.3 Mistrusting the Science?

It seems unlikely that the experience of extreme weather and the depictions that extend and sharpen such extremes into a bleak and cataclysmic future will have any substantive impact upon public perceptions of climate change and trust in the stability of climate. One tempting response might be to urge scientists to play a central role in communicating the sober and significant possibilities of a volatile climate. The problem here is that just as trust in the climate is under scrutiny, so is the science that does the scrutinising. In this section we consider the role of science in society and policymaking, and the fact that climate science cannot cordon itself off from the politics that both requires and undermines it.

For many, the reason for public apathy on the issue of climate change is their lack of scientific knowledge. When considered in this way, the solution might be seen simply as a matter of communicating and disseminating the science more widely or more transparently. The communication of scientific knowledge is indeed an important obstacle. Because climate change is not experienced directly, perceptions must be drawn from the complex data and often esoteric research generated by climate scientists, which is then translated by various actors and institutions: think-tanks, journalists and politicians

(Ruser, 2018). "On any complicated and controversial issue, non-expert citizens, including almost all of the public, face a challenge in discerning fact from fiction." (Minozzi, 2011: 301). Competitive political elites can "jam" messages from their opponents, exploiting uncertainty of the public by presenting them with directly contradictory information (Minozzi, 2011) and the sensationalism and bias of the media can hinder the communication and understanding of climate science (Cook, 2007). The public reception to the issue of climate seems to hinge upon political alignment, media interpretation and socio-economic values.

Indeed, numerous studies have shown that views regarding climate change and climate science and policy are divided often along political party lines. For example, a recent Pew Research Centre report on climate politics in the US diagnoses "major divides" in the way the American public interprets the science of climate change. It explains that the political left and right have "vastly divergent perceptions of modern scientific consensus, differing levels of trust in the information they get from professional researchers, and different views as to whether it is the quest for knowledge or the quest for professional advancement that drives climate scientists in their work." (Pew Research Centre, 2016). US citizens voting for the Democratic Party are more likely to believe in climate change and climate scientists, whereas Republican Party voters are less likely to expect severe harms from climate change and are more sceptical of the claims of climate scientists.

But the point we want to make here is not simply that public perceptions of climate change and climate science are not always entirely rational. It is rather that what is "rational" is precisely what is at stake in discussions of climate change. As we hope to have shown in Chapters Two and Three, measurements of climate and predictions regarding climate change are extremely difficult. Scientific uncertainty is inevitable. This means there is no one "rational" response, which makes policymaking extremely difficult and highly controversial. Decisions about the future must be made using scientific advice that cannot offer precise data or forecasts.

But even if there existed a perfect knowledge base about climate sensitivity to changing environmental conditions and climate impact, then tackling climate change would still not become a matter of technocratic management of ecological resources and socio-economic practices, for the non-rational behaviour and different cultural understandings of human beings make policymaking highly complex. The prognosis of the impacts of climate change, of the public response to a climate policy, and of the effects of a climate policy on our socio-economic system is consequently extremely difficult, if not impossible. One cannot simply project, given social value systems, techniques, structures and processes, into the future. This is particularly true in modern "knowledge societies", which are not only characterised by a rapid tempo of social change and a growing dependency upon scientific knowledge, but also by the growing incapacity of large institutions, like the government or science, to define and put into motion suitable, socially acceptable solutions for many occurring problems (Stehr, 2001).

Apathy towards climate change cannot be blamed upon a deficit of information, although many assume that this is the reason (Norgaard, 2011: 1–2). Even if the public had a better grasp on scientific procedures, instruments and data, the issue would still not be straightforward. The existence of a plurality of perspectives and values means that there is no one "right" path to follow in responding to climate change. Climate change is characterised as a "wicked issue" in which the consequences are uncertain, and the stakes are high (Hulme, 2009). Scientific knowledge indubitably and invaluably offers an important resource for policymakers. But if policy can be *informed* by science policy it cannot be *dictated* by it. Science can provide and interpret data, but it cannot ultimately tell us how to use that data. Science cannot tell us what priority to give climate change vis-à-vis other pressing social, economic and environmental issues. It cannot tell us how to allocate resources; whether to focus on climate change mitigation or adaptation, or to pursue geo-engineering or renewable energy or wide sweeping lifestyle change. These are political and ethical decisions that cannot simply be "read off" from scientific data.

The lack of perfect knowledge of climate change and the lack of consensus about what to do with such imperfect knowledge, is the reason why there is a call for the acknowledgement of the uncertainties and disagreements that surround both climate science and climate policies, rather than the assertion of scientific consensus on the "fact" of anthropogenic climate change (Machin, 2013). As one commentary asserts: "Building the basis for policy action cannot be done simply with appeals to fact. Where these facts are complex and negotiated, as in the case of climate change, experts and policymakers need to acknowledge and engage more actively with public 'matters of concern'." (Pearce *et al.*, 2017).

The danger here is that the assertion of certainty when there is none can be easily exploited by those unconvinced by the possibilities or urgency of climate change (Machin *et al.*, 2017). If levels of trust in climate science and scientists are in decline, then any scandal — true or false — can mean a backlash against climate policy. For example, the so-called "climategate" affair, refers to events occurring just weeks before the Copenhagen summit, when the email accounts of scientists working at the Climatic Research Unit (CRU) at East Anglia University in the UK were hacked. It was claimed, erroneously, that the email exchanges between scientists illustrated the fact that climate change was a conspiracy. What the affair illustrated, on the contrary, was the highly politicised atmosphere around the issue of climate change. As the British governmental report written in response to climategate states: "It is self-evident that the disclosure of the CRU e-mails has damaged the reputation of UK climate science, and as views on global warming have become polarised, any deviation from the highest scientific standards will be pounced on." (British Government, 2010).

This is why we submit that climate change cannot solely be a concern for natural scientists, since both the issue and the research are embedded in social and political processes. The task of climate and climate impact research cannot focus only upon natural climate variations and anthropogenic climate changes. Rather, the social sciences should contribute to the effort of understanding the complex problem of the perception of the climate and its changes. Without a robust multidisciplinary approach to inform policymakers, it might be feared that any adopted climate policy will be less consistent and less reliable than the climate itself. In the next chapter we consider some of the possible options for policymakers.

5.4 Conclusion: Social Change?

Will environmental conceptions sustained during centuries of increasing security from the hazardous and capricious nature of climatic conditions be disrupted by the increasing incidences of "extreme weather"? May nature be about to "remind us of its existence" and force us to think in the long term (Serres, 1995: 29)? There is no guarantee of this. Norgaard asks: "Given that many people do know the grim facts, how do they manage to produce an everyday reality in which this urgent social and ecological problem is invisible?" (Norgaard, 2011: xviii). Amitav Ghosh argues that it is because "the climate crisis threatens to unravel something deeper" (Ghosh, 2016: 138) that there is resistance to its reality. This chapter, we hope, has shown that trust in the stability of climate is not necessarily deposed by experience, nor by depictions of catastrophe, nor scientific data. This would be to undermine the extent to which many societies are underpinned and reproduce expectations about environmental conditions, including the climate. Perceptions of climate and climate change are shaped by social norms, political ideologies and cultural practices (Weber, 2010; Ghosh, 2016). Our understanding of the climate is not simply read off the weather, nor scientific data and predictions. The social meaning of climate is constructed through various expectations, values, ideas and emotions regarding nature and science. The question is whether this construction will change and if so, how quickly, for the possibilities of formulating and implementing new innovative and effective climate policies depend upon public perceptions. Bearing that in mind we turn to the options of climate policy in the next chapter.

References

Barber, Bernard (1983) *The Logic and the Limits of Trust*. New Brunswick, New Jersey: Rutgers University Press.

British Government (2010) *Government Response to the House of Commons Science and Technology Committee 8th Report of Session 2009–2010: The Disclosure of Climate Data from the Climatic Research Unit at the University of East Anglia*. London: The Stationary Office Ltd.

Brönnimann, Stefan (2015) *Climatic Changes since 1700*. Heidelberg: Springer.

Brückner Eduard (1890) "Climate Changes since 1700". In: *Eduard Brückner — The Sources and Consequences of Climate Change and Climate Variability in Historical Times*. (Eds.) Stehr, Nico and Hans von Storch. Springer, 2000.

Cook, Peter (2007) "Public science, society and the greenhouse gas debate". In: *Public Science in Liberal Democracy*. (Eds.) Porter, Jene and Peter Phillips. Toronto, Buffalo and London: University of Toronto Press, pp. 84–108.

Egan, Patrick J. and Megan Mullin (2012) "Turning personal experience into political attitudes: The effect of local weather on Americans' perceptions about global warming," *The Journal of Politics* **74**: 796–809.

Flannery, Tim (2009) *Now or Never: Why We Must Act Now to End Climate Change and Create a Sustainable Future*. New York: Atlantic Monthly Press.

Fukuyama, Francis (1995) *Trust: The Social Virtues and the Creation of Prosperity*. New York: Free Press.

Ghosh, Amitav (2016) *The Great Derangement: Climate Change and the Unthinkable*. London and Chicago: University of Chicago Press.

Giddens, Anthony (1990) *The Consequences of Modernity*. Stanford: Stanford University Press.

Giddens, Anthony (2009) *The Politics of Climate Change*. Cambridge, Malden. Polity Press.

Gore, Al (2006) *An Inconvenient Truth: The Planetary Emergency of Global Warming and What We Can Do About it*. London: Bloomsbury. Back cover.

Hansen, James, Makiko Sato and Reto Ruedy (2012) "Perception of climate change," *PNAS*. E2415–E2423.

Hood, Marlow (2017) "2020 deadline to avert climate catastrophe: Experts," June 28. Available at: https://phys.org/news/2017-06-deadline-avert-climate-catastrophe-experts.html

Hulme, Mike (2009) *Why We Disagree About Climate Change: Understanding Controversy, Inaction and Opportunity*. Cambridge: Cambridge University Press.

Kolbert, Elizabeth (2016) "Climate catastrophe, coming even sooner?" *The New Yorker*. 31 March. Available at: https://www.newyorker.com/news/daily-comment/climate-catastrophe-coming-even-sooner

Lovelock, James (2011) "Forword". In: Henson, *The Rough Guide to Climate Change*. London: Rough Guides.

Luhmann, Niklas (1979) *Trust and Power*. New York: Wiley.

Machin, Amanda (2013) *Negotiating Climate Change*. London: Zed Books.

Machin, Amanda, Alexander Ruser and Nikolaus von Andrian Werburg (2017) "The climate of post-truth populism: Climate vs the people," *Public Seminar*. Available at: http://www.publicseminar.org/2017/06/the-climate-of-post-truth-populism/#.WbJ3AGSCwfE

Marsh, George P. (1965) [1864, 1ˢᵗ ed.] *Man and Nature*. Cambridge, Mass.: Harvard University Press.

Minozzi, William (2011) "A jamming theory of politics," *The Journal of Politics* **73**: 301–315.

Moser, S. C. and Dilling, L. (2011) "Communicating climate change: Closing the science-action gap". In: *The Oxford Handbook of Climate Change and Society*. (Eds.) Dryzek, J., R. B. Norgaard and D. Schlosberg. Oxford: Oxford University Press, pp. 161–174.

Myers, Teresa A., Edward W. Maibach, Connie Roser-Renouf, Karen Akerlof and Anthony A. Leiserowitz (2012) "The relationship between personal experience and belief in the reality of global warming," *Nature Climate Change* **2**: 1–5.

Nerlich, Brigitte and Jaspal, Rusi (2014) "Images of extreme weather: Symbolising human responses to climate change," *Science as Culture* **23**(2): 253–276.

Newton, Kenneth (2007) "Social and political trust". In: *The Oxford Handbook of Political of Behavior* (Eds.) Dalton, Russel J. and Hans-Dieter Klingemann. Oxford: Oxford University Press, pp. 342–361.

Norgaard, Kari Marie (2011) *Living in Denial: Climate Change, Emotions and Everyday Life*. Cambridge, London: MIT Press.

Offe, Claus (1999) "How can we trust our fellow citizens". In: *Democracy and Trust*. (Ed.) Mark Warren. Cambridge: Cambridge University Press, pp. 42–87.

O'Neill, Onora (2002) *A Question of Trust*. Cambridge: Cambridge University Press.

Okasha, Samir (2016) *Philosophy of Science*. Oxford: Oxford University Press.

Pearce, Warren, Reiner Grundmann, Mike Hulme, Sujatha Raman, Eleanor Hadley Kershaw and Judith Tsouvalis (2017) "Beyond counting climate consensus," *Environmental Communication*.

Pew Research Centre (2016) "The politics of climate," Available at: www.pewinternet.org/2016/10/04/the-politics-of-climate/

Putnam, Robert D. (1995) "Bowling alone: America's declining social capital," *Journal of Democracy*, **6**(1): 65–78.

Revill, Chloe and Victoria Harris (2017) *2020: The Climate Turning Point. Mission 2020*. Available at: http://www.mission2020.global/

Ruser, Alexander (2018) *Climate Politics and the Impact of Think Tanks: Scientific Expertise in Germany and the US*. London: Palgrave Macmillan.

Scranton, Roy (2018) *We are Doomed. Now What? Essays on War and Climate Change*. New York: Soho.

Sen, Amartya (1999) *Development as Freedom*. New York: Alfred A. Knopf.

Serres, Michel (1995) [1992, 1ˢᵗ ed] *The Natural Contract*. Ann Arbor: The University of Michigan Press.

Shardlow, Matt (2016) "Nature skips a season," *The Guardian* January 1. Available at: http://www.theguardian.com/environment/2016/jan/01/nature-skips-season

Simmel, Georg (1906) "The sociology of secrecy and of secret societies," *American Journal of Sociology* **11**: 441–498.

Stehr, Nico (2001) *The Fragility of Modern Societies: Knowledge and Risks in the Information Age*. Sage: London.

Stehr, Nico and Hans von Storch (2000) "Eduard Brückner's ideas — Relevant in his time and today". In: *Eduard Brückner — The Sources and Consequences of Climate Change and Climate Variability in Historical Times*. (Eds.) Stehr, Nico and Hans von Storch. Springer.

Stehr, Nico, Hans von Storch and Moritz Flügel (1995) "The 19th century discussion of climate variability and climate change: Analogies for the present?" *World Resources Review* **7**: 589–604.

Swyngedouw, Erik (2011) "Whose environment? The end of nature, climate change and the process of post-politicisation," *Ambiente and Sociedade* **14**(2): 69–87.

Weber, Elke U. (2010) "What shapes perceptions of climate change?" *Climate Change* **77**: 332–342.

CHAPTER 6

Climate as Policy Issue

Clear waters and green mountains are as good as mountains of gold and silver.

Xi Jinping, President of the People's Republic of China,
Address to the United Nations, 18 January 2017[1]

As the previous chapters have hopefully demonstrated, climate is no longer simply just a background condition of everyday life, whether it be a constant or capricious one. Today, climate is something that demands active assessment and response. The changing climate, along with its concomitant consequences and risks, is now incorporated into our political institutions and mechanisms as an important issue for which policy should be determined and implemented. Climate is not something that is simply *external* to human society — it is something that is *internal* to its politics, perhaps even occasionally occupying centre stage. Climate is something that policy is made *for* and *about*.

In this chapter we consider the various ways in which the changing climate is understood as a policy issue. We distinguish three distinct, but not mutually exclusive, approaches to climate policy. All these approaches are concerned either with adapting to a capricious climate, which is being definitely, albeit unpredictably, changed by human society, or by mitigating those uncertain changes.

First, there is an approach to climate change that regards it as a matter of *scientific or technological policy*. Science and scientists have tended to dominate climate policy discussions and are often expected to deliver the solutions. Technologies are considered key to

---

[1] See https://www.theguardian.com/world/2017/jan/19/chinas-xi-jinping-says-world-must-implement-paris-climate-deal

both climate change adaptation and mitigation. These sorts of policies emphasise the importance of technological innovations, such as a shift to nuclear power, the development of machinery able to harness energy from renewable sources, the implementation of carbon capture and storage and geo- or climate-engineering. Second, climate change is seen as falling within the remit of *economic policy*. In this set of policies, behaviour is "managed" through market mechanisms such as emission trading and taxation, subsidy of renewable energy sector and consumer "nudging". Finally, a different policy approach is that of *regulatory policy*. Regulatory mechanisms are used to propel a shift across society and industry, towards more sustainable patterns of energy consumption. These sorts of policies implement rules on recycling, laws on land use and standards on building materials.

We argue that no single set of policy solutions is likely to produce a perfect "solution" to a changing climate; no one policy instrument is able to mitigate climate change or to stream-line adaptation. Rather, we suggest that climate policy needs to be regarded not as a separate, but rather as significantly interconnected with other policy *areas*, such as energy, transport, security and health, for example. We also highlight the way in which climate change is enmeshed with other policy *issues*, such as poverty, inequality and security. It is partly because of the complex embeddedness of climate policy that we argue that it is determined by democratic mechanisms. The final section of the chapter considers whether, as appears to be commonly believed, robust climate policy and democracy are incompatible.

6.1 Scientific Policy

As has been argued in this book, climate and climate change are often regarded as primarily — sometimes even solely — a scientific matter. Scientists are expected to accurately depict the situation, forecast the future and delineate the environmental limits within which society can sustainably function. In addition to this, they are expected to produce the technological innovations that deflect the dangers of a changing climate.

In a survey of these technological innovations, this category can itself be subdivided into three categories: (1) *defensive* technologies that promise to protect human beings and their social and natural environment against climate extremes, (2) *alternative* technologies that proffer cleaner forms of energy production in order to reduce social dependency on fossil fuels and greenhouse gas emissions, and (3) *engineering* technologies that offer to diminish the impact and severity of a warming planet by "managing" the climate.

Technological innovations have always been crucial in allowing human beings to protect themselves and their resources from the hazards of extreme weather. An important and perhaps easily forgotten example is the invention of air-conditioning that produced huge migration flows (Biddle, 2008).[2] The utilisation of residential and office air conditioning is expected to increase dramatically over the next several decades as global temperatures rise along with incomes. Other examples include flood defences, drainage systems, hurricane resistant architecture, drought tolerant crops, water-saving "thirsty houses" (Weiner, 2015), skyscraper pendulums (McKean, 2013) and vaccinations. Alongside new "high" technologies are more traditional schemes and methods used and refined over generations (UNEP, 2012). The utilisation of these sorts of technologies implements a defensive response, in reaction to as well as in anticipation of a changing climate that exacerbates existing environmental hazards.

Alongside these are technologies that attempt to tackle one of the *causes* of such threats, by reducing the release of greenhouse gases into the atmosphere. Alternative technologies attempt to harness cleaner and renewable sources of energy. Equipment to generate energy from wind, solar thermal, photovoltaic, geothermal, tidal and hydropower energy, for example, are currently in various stages of development. There are also technologies that enhance energy efficiency, for example

---

[2] As Jeff Biddle (2008: 412) notes in the United States: "In 1955 fewer than 2 per cent of the nation's residences had air conditioning; by 1980 over half were air conditioned.

software that allows farmers to determine when and how to apply to nitrogen to their crops, which can reduce the amount lost into the atmosphere. As a greenhouse gas, nitrogen is 300 times as potent as carbon dioxide (Burgess, 2014).

Another emerging technology is that of carbon sequestration and storage (CCS), in which carbon dioxide is captured out of exhaust streams and held in non-atmospheric reservoirs.[3] This technique could potentially help stem the emission of carbon dioxide particularly from coal power plants, which according to a UNEP report are the source of over a quarter of worldwide greenhouse gas emissions (UNEP, 2012). The methods of CCS vary; carbon can be captured most easily post-combustion, by injecting a chemical into the waste, but this lowers energy production and massively increases costs. Pre-combustion is a newer technique but can only be implemented on new "supercritical" coal plants that are also extremely expensive. The other issues, aside from the costs include the question of where to store the captured carbon dioxide and how to get it there. Proponents suggest that it can be injected in liquidised form in deep underground storage. But not everyone is convinced that this is safe. Moreover, tens of thousands of new pipelines would be needed to move the gas around (Zeller, 2013).

Even more controversially, nuclear fission also offers cleaner energy. Nuclear power produces virtually no greenhouse gas emissions and the uranium fuel that it uses is in abundant supply. But in contrast to coal and natural gas power stations, while the fuel is relatively plentiful and cheap, the nuclear power plant is expensive — it involves enormous up-front costs of up to 10 billion US dollars (Findlay, 2010). It also takes a long time (on average seven years) to plan, win regulatory approval and build such a plant (Findlay, 2010: 15). The International Atomic Energy Agency (IAEA) states that a nuclear power programme demands a commitment of 100 years.[4] Perhaps the biggest constraint on pro-nuclear energy policy is the ineradicable risk involved and the concomitant anaemic and erratic levels of public support. Nuclear

---

[3] See the MIT CC&ST Program website at: https://sequestration.mit.edu/

[4] See the International Atomic Energy Agency website at www.iaea.org

energy is highly contested by those concerned by the environmental risk entailed, however alledgedly small it might be. One area of concern is that of the proper (and sustainable) disposal of radioactive waste.

A typical nuclear reactor will produce 27 tonnes of used fuel a year produced by burning uranium. This used fuel contains radioactive "high-level waste", which must be carefully managed. The World Nuclear Association explains that the waste can be safely stored in ponds, disposed of, or reprocessed (World Nuclear Association, 2016). If it is reprocessed the extracted uranium and plutonium can be recycled and the amount of waste is reduced to 3 m^3 each year. They also point out that after 40–50 years the radioactivity falls to one thousandth of that level. Yet concern about nuclear waste acts nevertheless as a "dampener" on the industry (Findlay, 2010). There are also the problems of regulation. Although there are various conventions and regimes, including the IAEA in Vienna and the OECD in Paris, they are not necessarily well integrated. Moreover, since each country is ultimately responsible for the safety of its nuclear industry, there is a danger that "new entrants will be unaware of and unprepared for their safety responsibilities, have no safety culture and be too poorly governed to enforce safety regulations" (Findlay, 2010: 23). Even states that have been generating nuclear energy for decades, however, are not immune to accidents. The 2011 disaster of Fukushima provoked lingering concern about the risks that are taken in pursuing such technology, and there is also the worry that nuclear power plants could be targets of terrorist attacks. For Greenpeace, an environmental NGO campaigning against nuclear power, for example: "None of the world's 436 nuclear reactors are immune to human errors, natural disasters, or any of the many other serious incidents that could cause a disaster. Millions of people who live near nuclear reactors are at risk".[5] Proponents of the nuclear option argue precisely the opposite. A letter written by a group of scientists, including James Hansen, states that the risk lies rather in failing to utilise the potentials of this form of energy:

---

[5] See the Greenpeace campaign on nuclear and its reports on the Fukushima Nuclear Disaster at http://www.greenpeace.org/international/en/campaigns/nuclear/Fukushima-disaster/

"Continued opposition to nuclear power threatens humanity's ability to avoid dangerous climate change" (CNN, 2013). Nuclear power, thus, is simultaneously regarded both as the best means of addressing climate change... and the worst.

The emergence of "geo-engineering" or "climate engineering" is also currently emerging as a serious option in climate policy discussions (Hamilton, 2014; Hulme, 2014). Geo-engineering can be defined as "the deliberate large-scale intervention in the earth's climate system" (The Royal Society, 2009: IX). David Keith, a professor of applied physics and social policy, is an advocate of such deliberate intervention in the climate, arguing for a form of "solar geo-engineering" in which the injection of reflective particles of sulphuric acid creates a "sunshade" for the earth below. He claims that "the underlying science is sound, and the technological developments are real" and therefore such solar geo-engineering is "a cheap tool that could green the world" (Keith, 2013: X). To be sure, Keith admits that the use of such technology is a hard choice; it is an "extraordinarily powerful tool yet it is also dangerous" (Keith, 2013: X). Still, he offers a qualified endorsement: "If research shows that these technologies have benefits that greatly exceed the costs, then we should in fact start relatively soon, albeit carefully and with small steps." (Grolle, 2013).

Scientists themselves warn, however, that no particular technology is able to provide a straightforward "silver bullet" (The Royal Society, 2009: V). Indeed, these technologies may well generate their own damaging impact and aggravate in complex and uncertain ways the delicate mechanisms affecting the environment. As an editorial in *Nature* recently states: "The cure could be worse than the disease." (Nature, 2014: 8). The legitimacy and efficacy of technological scientific policies is highly contested, and thus their deployment must negotiate ethical questions and political contestation (Szerszynski and Galarraga, 2013: 2818). The Royal Society, in its widely cited report on geo-engineering, notes that while there are many *scientific* difficulties and uncertainties, "the greatest challenges to the successful deployment of geo-engineering may be *social, ethical, legal and political* issues associated with governance, rather than scientific and technical issues" (The Royal Society, 2009: XI).

Indeed, the possibilities of all these forms of technologies hinges, of course, upon whether governments are prepared to subsidise and support their development and implementation. These policy decisions hinge upon factors such as ethical legitimacy, energy security, public support and economic viability. It is this latter concern that we turn to next. The economic policy approach suggests that economic viability is crucial to an efficacious climate policy.

6.2 Economic Policy

One important category of policies is the one encompassing policies that attempt to implement behaviour change in individuals and industries by way of monetary incentives. The aim here is to align environmental goals with economic motivations. Through various financial schemes, strategies and subsidies, society is prodded into more environmentally friendly behaviour.

This approach to climate policy is supported by the discourse of "ecological modernisation", in which economic growth and environmental protection are not seen as antagonistic to each other, but rather as mutually reinforcing (Berger *et al.*, 2001 ; Machin 2019). The "win-win storyline" of this model asserts that it is economically rational to combat climate change, since it is good for business (Hillebrand, 2013). And this is exactly the promise made by various politicians. For example, at the 2014 UN climate summit, (then) US President Barack Obama stated: "… there does not have to be a conflict between a sound environment and strong economic growth." (Obama, 2014). Former UK Prime Minister David Cameron made an identical claim in the same year: "There need not be a trade-off between economic growth and reducing carbon emissions … We need to give business the certainty it needs to invest in low carbon … It means championing green free trade, slashing tariffs on things like solar panels." (Cameron, 2014). Similarly, in the EU *Climate Action* document, the claim is reiterated: "Early action to develop a low-carbon economy is helping to boost jobs and growth" (EU, 2013).

The absence of any contradiction between economic and environmental priorities was supported by the 2006 publication of the

Stern Review, commissioned by the British government and written by a team headed by economist Nicholas Stern. The review offers a cost-benefit analysis suggesting that the economic benefits in tackling climate change outweigh the costs, calculating the cost of action at 1 per cent of global GDP each year and the cost of *not* acting to be equivalent to 5 per cent of global GDP each year: "[T]he evidence gathered by the Review leads to a simple conclusion: the benefits of strong early action considerably outweigh the costs. Ignoring climate change will eventually damage economic growth." (Stern, 2006: ii).

As the Stern Review acknowledges, however, government involvement is needed to force the market to incorporate environmental costs that are currently "externalised". The environmental impacts of industrial production currently do not feature in economic calculations and the generation of prices. When a factory pollutes a river, for example, the impact this has on the wider eco-system (including the human communities who depend upon it) is not incorporated into its costs. Similarly, the huge impact to the climate from greenhouses gas emissions is not borne by those who make profits from the processes that generate the emissions. Thus, as Stern concludes, in a widely cited statement, climate change is "the biggest market failure ever seen" (Stern, 2006: i). As one analysis puts it: "[M]arkets... generate prices that do not reflect the true cost to society of our economic activities; these prices do not give the correct signals about where to devote resources." (Bownen and Rydge, 2011: 72). Many of those concerned by climate change thus argue that the environmental externalities of current market practices need to be internalised through government policies that compel polluters to pay.

One method of forcing industries and energy companies to internalise environmental costs consists of "putting a price on carbon". The assumption here is that adding a monetary cost to the carbon that is burnt and released as a gas into the atmosphere, will incentivise both greater energy efficiency and the development of green technologies. A major policy debate concerns the best way to do this. The debate mainly revolves around two options: carbon taxing and carbon trading (Stavins, 2008).

Carbon taxes (or when other greenhouse gases are included, "emissions taxes") have been introduced in the Scandinavian countries in the early 1990s and more recently in the Netherlands, Slovenia, Germany, the UK and Ireland (Senit, 2012). Carbon taxes are a "Pigouvian" form of tax because they are not primarily aimed at raising money but rather at correcting the problem of negative externalities. The tax may be targeted at either "up-stream" or "down-stream" polluters, but they should motivate polluters to reduce their emissions by either saving energy or switching to new forms of energy. They are nonetheless promoted as having a "double dividend"; not only do they provide an unambiguous signal to economic agents to reduce carbon consumption but they also produce revenue that can be spent upon environmental measures, or upon making the tax "fiscally neutral" by reducing other taxes (Metcalf, 2009; Pearce, 1991). This might be important because taxes on energy consumption disproportionately effect the poor. Deciding the level of the tax, however, is not a straightforward matter. Another problem is that a tax cannot guarantee overall emissions reductions; while it might incentivise greater energy efficiency, through the development of green technology and energy saving, this might not achieve absolute reductions (Hillebrand, 2013). Moreover, any form of tax is regarded as unpopular with voters, for whom it is believed that "tax" is "almost a four-letter word" (Nordhaus, 2007). This in turn can make them unattractive to politicians (Green Fiscal Commission, 2009).

The policy of carbon (or emissions) trading is often preferred by economists, who argue that it allows flexibility for achieving the required reductions in emissions, and that it is more cost-effective than taxation (Stavins, 2008). Over the last two decades, carbon markets have consolidated and are widely regarded as a key policy instrument in tackling climate change (Stephan and Paterson, 2012). These markets involve the trading of carbon credits that are distributed to various industry sectors and then capped at a certain level. If a particular industry or business needs more, they must buy them on the market. The incentive is thus to reduce the demand for credits and lower the emission of greenhouse gases. The largest emissions market is the

Emission Trading Scheme of the European Union (EU-ETS) introduced in 2005 and now in its "third phase". The EU-ETS is a cornerstone in the EU's climate policy and covers just under half of the EU's greenhouse gas emissions (Convery, 2009). The aim is to reduce the emissions by 43 per cent by 2030 (starting from 2005), from the sectors it covers. The scheme has recently been revised so that the overall quantity of allowances is reduced by 2.2 per cent every year from 2021 (European Commission, 2015).

As some commentators have pointed out, it is simply taken for granted that emissions markets make a valuable contribution (Stephan and Paterson, 2012). And yet important problems arise with this policy instrument: carbon is subject to price fluctuations, it does not offer such a clear price on emissions as a tax, and it only targets some industries and sectors (Schuppert, 2011; Gilbertson and Reyes, 2009). Others argue that it has been "gamed" by corporate lobbyists who have demanded more credits than they need and thus ensured a surplus that they are able to profit from (Ball, 2014). The fact that the credits are awarded to industries according to their previous volumes of emissions (a process known as "grandfathering") leads some critics to point out that "those who have polluted most in the past are rewarded with the greatest subsidy… a free gift of pollution rights to some of the worst industrial polluters" (Gilbertson and Reyes, 2009: 10).

Another problem has appeared with the connection of the emissions trading system with the Clean Development Mechanism (CDM) introduced by the Kyoto Protocol in 1997 (UNFCCC). The CDM allows wealthy industrialised countries to offset their emissions by investing in carbon saving in the developing world. For example, a solar-power farm in China or a project to slow deforestation in Brazil, which indirectly reduces carbon emissions, can then apply for certified emission reduction (CER) credits. These credits can be sold onto a country that is part of the EU-ETS. It is usually easier to achieve emission reduction targets by investing in new technology than to change already existing infrastructure (Wara, 2007). Linked together in this way, the CDM and the EU-ETS are "the keystones of an emerging global regime of linked but distinct markets for greenhouse-gas emission-control" (Wara, 2007: 595). However, there are various flaws. Firstly, this has enabled developed countries to defer any domestic structural change. As one critical paper states: "Trading has been used by industrialised nations to dodge some of their own domestic CO_2 reduction commitments in exchange for financial contributions to cuts beyond their own borders." (Moreno *et al.*, 2015). Secondly, many of the projects approved by UN officials in the past were not legitimate and some even contributed to other environmental problems. For example, the creation of large plantations of trees generates credits. But in order to make space for such large plantations, habitats that had contained a rich diversity of plants and animals were destroyed (Ball, 2014). Thirdly, various projects that may have been carried out anyway have been included under the CDM, thus diverting resources that could have been more usefully invested in other projects (Wara, 2007). Finally, offsets are extremely difficult to measure: "Since carbon offsets replace a requirement to verify emissions reductions in one location with a set of stories about what would have happened in an imagined future elsewhere, the net result tends to be an increase in greenhouse gas emissions." (Gilbertson and Reyes, 2009: 11).

The debate between taxing on the one hand and trading on the other has tended to dominate policy discussions and can be accused of crowding out other policy options. Other important economic policy instruments that may be helpful in tackling climate change exist. For example, the strategic use of subsidies to encourage entrepreneurs into research and development of low carbon energy technologies, and the creation of congestion charge zones to encourage individuals to use public transport.

However, all these policies arguably suffer from the problematic assumption that if governments put the right economic signals in place, for example by "putting a price on carbon", the market will do the work (Stern, 2009: 99). The reality is not so straightforward. This policy approach reduces individuals to economic agents who are solely incentivised by the "right prices" (Stern, 2009: 100). It forgets that individuals are not only consumers and producers, but human beings with various ideas, interests, identities and values whose actions do not necessarily fit with the predicted "rational behaviour" (Stephan and Paterson, 2012). As Dale Jamieson notices: "People often act in ways that are contrary to what we might predict on narrowly economic grounds." (Jamieson, 1992: 144). Behaviour that does not conform to economic self-interest cannot be dismissed as a result of lack of information, certainty or transparency; human beings invariably and inevitably act in non-rational ways for non-rational reasons: "People are creatures of social routine and habit, and of fashion and fad. These patterns of routine and fashion stem from how people are locked into and reproduce many different kinds of social institutions, both old and new." (Szerszynski and Urry, 2010: 3). This is not to say that routines and fashions cannot change or be changed, but that this is often not entirely economically "rational".

It cannot be assumed that society is simplistically moulded through economic policy and that economic policy is enough to "green" society, but this is often forgotten by advocates of an economic approach to tackling climate change (Machin, 2013: 16–27). Indeed, climate change is commonly constructed not only as a market failure but as a "market opportunity" too, from which budding entrepreneurs make profits from climate friendly innovations and everyone stands to benefit. But this belies the broader issue of structural change; rather than changing the dominant order of things, these sorts of economic solutions can *reinforce* it. As Richard York argues, tackling the environmental crisis may demand "taking actions that challenge the growth paradigm that is fundamental to the modernisation project", rather than reaffirming it (York, 2010: 8).

Economic policies may well be crucial in underpinning a shift to a less environmentally destructive world, but they can only work if they are implemented alongside *other* policies (Green Fiscal Commission, 2009; Janicke, 2004). While the dominant framing of climate change policymaking assumes that environmental goals can be reconciled with economic ones, this ignores the possibility that there might actually be a serious tension between the two (Berger *et al.*, 2001). It might well be the case that effective climate policy demands more radical policies (Moreno *et al.*, 2015). It is also possible that for affective climate policy, economic instruments need to be complemented by a regulatory approach.

6.3 Regulatory Policy

The final category of policies we assess here are those that restrict environmental damage through rules and regulations. Through, for example, standards and laws regarding building practices and material, pollution levels, deforestation, land use or regulations to leave fossil fuels unused. This policy approach acknowledges that the right energy-saving technologies may not appear as quickly as hoped, and that markets may not work as rationally and efficiently as expected.

An example of regulatory policy is the creation, revision and implementation of laws regarding land use in the Brazilian Amazon. Brazil is home to 64 per cent of the world's rainforest, but over 700,000 square kilometres (an area twice as large as Germany) has been cleared since 1970 (Yale Forest Atlas a, n.d.). Not only does deforestation reduce biodiversity, destroy the homes of native peoples and the habitats of numerous species, and damage an important resource for regulating the earth's climate — the "lungs of the world" — it also releases tonnes of carbon stored in trees into the atmosphere. Forests are cleared mainly for agricultural reasons; pasture for beef cattle and crops such as soybeans. Deforestation is possibly the second-largest source of carbon emissions after fossil fuel combustion (van der Werf *et al.*, 2009). There is concern, moreover, that the deforestation of the Amazon basin could reach a tipping point, where the ecosystem is no longer able to regulate itself and shuts down, a situation known as "Amazon dieback" (Tollefson, 2013).

However, the general reduction of deforestation rates in Brazil by 70 per cent below the average of 1996 to 2005 has made Brazil "the global leader in climate change mitigation" (Nepstad *et al.*, 2014) and a potential model for other states. As Jeff Tollefson writes: "If other countries follow suit by protecting and expanding forests, which lock carbon up in trees and soils, they could slow the growth of global CO_2 emissions." (Tollefson, 2015). But this success is attributed to different causes: "The story varies depending on who is telling it." (Tollefson, 2015). Some argue that market forces play the most important role (Nepstad *et al.*, 2014). Environmentalist campaigns may also have played a part (Tollefson, 2015). Greenpeace, for example, has run several campaigns putting pressure on the soybean and cattle industries (Greenpeace, 2009).

Yet government enforcement seems particularly crucial (Nolte *et al.*, 2013). Brazil's "Forest Code", written in 1965, introduced protective measures for rainforests — forest landowners have to maintain 80 per cent of their property under forest cover (Yale Forest Atlas b, n.d.). However, Brazilian authorities have struggled to enforce the Forest Code; regulations require institutions and mechanisms with the capacity to enforce them. Reforms made in 2006 under President Luiz Inacio Lula da Silva (popularly known as "Lula") were made to both the laws and the infrastructure in an attempt to improve forest management and protection (Bauch *et al.*, 2009). So, while policy is made at a federal level, management and enforcement have been decentralised, which has both advantages and challenges (Bauch *et al.*, 2009). In 2009, a land registry was set up to help the Brazilian government to formally license and monitor agricultural operations (Tollefson, 2015).

Many believe the apparently positive results have largely come as a result of the UN REDD+ programme (Reducing Emissions from Deforestation and Forest Degradation), a scheme linked to the Clean Development Mechanism. In connection with the scheme, President Lula set up the "Amazon Fund" to help slow deforestation (Doyle and Lawrence, 2015). The fund consists of voluntary donations from various states, most notably Norway from where a total of $1 billion was donated between 2008 and 2015. This money is used to fund projects such as reforestation, the protection of springs and monitoring of the rainforest through satellite images.[6]

It seems that it is a combination of different policy approaches that has led to a reduction of deforestation (and even an increase in

---

[6] See the Amazon Fund website at http://www.amazonfund.gov.br/Fundo-Amazonia/fam/site_en

reforestation). However, it appears equally likely that without more structural change and the interconnection of different policy issues, such as inequality, slowing deforestation will remain a precarious feat: "Enforcement has increased, but the basic factors driving deforestation — including poverty and the profitability of agricultural land — have not changed" (Tollefson, 2015).

A regulatory strategy was, until recently, being implemented by the US Environment Protection Agency, which had begun an attempt to put in place federal restrictions on power-plant carbon emissions. This came as part of then President Barack Obama's "Clean Power Plan", which hoped to cut carbon dioxide emissions from power stations by nearly a third. Individual states would be responsible for meeting target reductions by 2022 (Harder *et al.*, 2015). On the one hand, these set clear targets regarding climate change. On the other, precisely because they are so visible, they are often resisted by the industry and private sector. The regulatory policy approach's greatest challenge is that such rules, which come without concomitant immediate financial benefits, are often seen as inevitably unpopular with the public. If voters are unlikely to support such measures, then politicians may well be unlikely to do so too. This was certainly seen to be the case with President Donald Trump who in March 2017 signed an executive order demanding the revision of the Clean Power Plan, as part of a general attempt to rollback Obama's environmental policies (Meyer, 2017; Davenport and Rubin, 2017). This leads us to the question of the apparent contradiction between democratic governance and climate change policy: Are these irreconcilable?

6.4 Climate Policy: Who Decides?

Many who are concerned about climate change are worried about the apparent mismatch between the knowledge of scientists and the urgency of climate change on the one hand, and the institutions of democracy on the other. Modern democratic governments have so far not only failed to shrink the gap between climate science and climate policy, but they appear to have actually gotten in the way. It might seem, then, that decisions should be taken out of the fumbling hands of the public and the officials they elect and placed in the steadier hands of experts. Al Gore's famous "inconvenient truth" seemingly demands the complements of an "inconvenient democracy" (Stehr, 2015). Might a "benign dictatorship" or "eco-authoritarianism" be more effective in addressing such global environmental problems (Machin, 2018)?

In the past, warlike conditions and major disasters typically were seen to justify the abolition of democratic rights, if only temporarily. The term "exceptional circumstances" refers to conditions often invoked to grant governments additional powers to avert unforeseen threats. For example, well-known climate researcher James Hansen summarised the general frustration when he asserted: "[T]he democratic process does not work." (Adam, 2009). James Lovelock, another long-time scientific voice of warning, compares climate change to war, emphasising that we need to abandon democracy in order to pull the world out of its state of lethargy (Lovelock, 2009). Dale Jamieson, too, exemplifies the sceptical view about the obstacles faced by our present political order in coping with the consequences of global warming. He warns that climate change presents us "with the largest collective action problem that humanity has ever faced… we have not designed political institutions that are conducive to solving them." He adds: "Sadly, it is not entirely clear that democracy is up to the challenge of climate change." (Jamieson, 2014: 16).

There is actually nothing new, as David Runciman has documented, in this "outburst of disgust with the workings of democracy" (Runciman, 2013). He points out that concern with the failure of democracy is a perennial feature of democratic life. In the 1970s a similar kind of scepticism arose around the question of the limits to growth and the survival of humankind. Scientists warned about the essential slowness and inflexibility of democratic institutions and expressed their preference for authoritarian solutions. Dennis Meadows (the co-author of the original *Limits to Growth*) reiterated more recently his profound concern about the limitations of democratic governance: "Democracy contributes nothing at the moment to our survival." (Himmelfreundpointer, 2012).

The argument about an "inconvenient democracy" derives its intellectual sustenance from a range of considerations: the limitations of human thought and its orientation towards short-term self-interest, the inability of government (given constitutional constraints) to attend to long-term goals, the influence of vested interests on the political agendas of the day, the widespread social addiction to fossil fuel, and the inability of the public and politicians to grasp scientific findings. There is indeed evidence that the public in the United States, for example, does not view climate change as an urgent threat that calls for significant behaviour change (Leiserowitz, 2006). It might seem then that democratic governance is inherently flawed as a method of tackling environmental problems.

Until recently, explicit expression of doubt about the virtues of democracy has been rare among intellectuals and politicians. But today, disenchantment is becoming even more vocal as entrenched climate policy regimes such as the United Nations Framework Convention fail to live up to their promise, and international conferences on climate change fail to achieve a substantial global reduction of greenhouse gas emissions. Leading climate scientists insist that humanity is at a crossroads. A continuation of present economic and political trends leads to disaster, if not a collapse of human civilisation. Should climate scientists then play a bigger role in deliberations regarding climate policy?

The idea that science and scientific leadership offer some sort of alternative to democracy has major weaknesses. Scientific knowledge cannot dictate what to do. Political decisions involve ethical and economic considerations too. Knowledge about how such decisions are best made is not particularly available to scientists. Indeed, such knowledge is inherently and necessarily contestable and even if the key uncertainties about climatic processes and risks were eliminated, the uncertainties associated with the social and political processes for taking effective action remain.

Nevertheless, accounts championing the ecological leadership of China currently abound (Stern *et al.*, 2017). In 2013 China declared a war on pollution (Huang, 2018) and has worked towards the establishment of an "eco-civilisation" (Kitagawa, 2017). As the supreme political authority of a non-democratic state, the Chinese Communist Party is able to implement efficient and enduring measures without the hindrance of democratic procedures and institutions. This is why Mark Beeson argues that authoritarianism is likely to be consolidated in regions that are particularly vulnerable to climate change and already have a propensity towards that form of regime: "[F]orms of 'good' authoritarianism, in which environmentally unsustainable forms of behaviour are simply forbidden, may become not only justifiable, but essential for the survival of humanity in anything approaching a civilised form." (Beeson, 2010: 289).

There is no consensus on the benefits of an authoritarian approach to environmental policy; however, empirical research actually suggests that democracy roughly correlates with a decrease in environmental degradation (Li, 2006; Neumayer, 2002). There might be various reasons for this. As Rodger Payne writes: "Thanks to free speech, free press and other individual liberties, it is possible for the combined forces of the mass media, various environmentalist movements and relevant scientific communities to monitor the activities of the most prominent sources of environmental degradation — often corporations and governments — and publicise their findings, however critical." (Payne, 1995: 44).

In any case, if we value democracy, it is crucial that we attempt to defend it against the threat of eco-authoritarianism. The defence of democracy in the face of climate change is difficult to maintain on the grounds that it produces the "right" results. But to believe that democracy is measured by the degree it guarantees the "right" policies, is to misunderstand democracy. As Tim Hayward points out, "a democracy is not merely a mechanism for electing governments but more importantly a type of society" (Hayward, n.d.). This means that democracy is not just the aggregative bargaining between different interests; life in a democratic society does more than permit the expression of predetermined and fixed demands — it foments identifications, ideas, movements and virtues that could potentially underpin the creation and implementation of climate policy and highlight unwanted side-effects, overlooked by centralised bureaucrats (Huang, 2018).

Ultimately, there can be no guarantee that democratic politics will produce sustainable policies, but democratic governments' ability to learn allows them, as Runciman explains, "to keep experimenting and adapting to the challenge they encounter, so that no danger becomes overwhelming" (Runciman, 2013: 319). It is only through revitalised democratic interaction that alternative perspectives that may provide new insights and ideas can be presented, tested and contested. Democrats can point to the rich variety of perspectives that provoke informed and creative environmental policymaking. But much alternative knowledge and ideas would be left untapped in a technocracy. An authoritarian or technocratic form of government has exactly the opposite effect, narrowing the complexity of social and political life precluding radical alternative perspectives. Furthermore, environmental policies made democratically, have a claim to far more legitimacy and are more likely to gain popular support than policies made by authoritarian governments who ultimately cannot be held accountable (Machin, 2018).

Perhaps our political institutions are failing us in the face of environmental threats today. But this is not because they are *too* democratic. Rather, it is because they are not democratic enough. Much of what counts as "democracy" today would hardly be promoted by democrats. The preferable option to the abolition or marginalisation of democracy, then, is its enhancement through the opening up of environmental politics to the diverse perspectives of individuals and groups offering novel, radical and often incompatible real alternatives to the status quo. Through the emergence of ideas and perspectives that may complement each other or compete for dominance, climate change can be connected to local and regional issues. It can be harnessed to, indeed perhaps even strengthen, local and regional capacities to respond to climate change. Broad participation of diverse voices and the encouragement of creativity and experimentation in the pursuit of multiple desired goals are crucial. For those who think that there can be only one global pathway in addressing climate change, the erosion of democracy might seem to be "convenient". But climate policy, we argue, must be compatible with democracy, otherwise the threat to civilisation will be much greater than that posed by the direct impacts of global warming.

References

Adam, David (2009) "Leading climate scientist: democratic process isn't working," *The Guardian*. Available at: http://www.theguardian.com/science/2009/mar/18/nasa-climate-change-james-hansen

Ball, Jeffrey (2014) "Facing the truth about climate change," *New Republic*. Available at: https://newrepublic.com/article/120914/how-congress-can-compromise-climate-change-legislation

Bauch, Simone, Erin Sills, Luiz Carlos Estraviz Rodriquez, Kathleen McGinley and Frederick Cubbage (2009) "Forest policy reform in Brazil," *International Forestry* **107**(3): 132–138.

Beeson, Mark (2010) "The coming of environmental authoritarianism," *Environmental Politics* **19**(2): 276–294.

Berger, Gerald, Andrew Flynn, Frances Hines and Richard Johns (2001) "Ecological modernisation as a basis for environmental policy: Current environmental discourse and policy and the implications of environmental supply chain management," *Innovation* **14**(1): 55–72.

Biddle, Jeff (2008) "Explaining the spread of residential air conditioning, 1955–1980," *Explorations in Economic History* **45**: 402–423.

Bownen, Alex and James Rydge (2011) "The economics of climate change". In: *The Governance of Climate Change*: *Science, Economics, Politics and Ethics*. (Eds.) Held, David, Angus Hervery and Marika Theros. Cambridge, Malden: Polity, pp. 68–88.

Burgess, Kaylegh (2014) "New technology helps farmers conserve fertilizer and protect their crops," *The Guardian*. Available at: http://www.theguardian.com/sustainable-business/2014/aug/20/adapt-n-harnessing-nitrogen-technology-farming-future

Cameron, David (2014) Speech to UN Climate Summit. Available at: https://www.gov.uk/government/speeches/un-climate-summit-2014-david-camerons-remarks

CNN (2013) "Top climate change scientists' letter to policy influencers," *CNN*. Available at: http://edition.cnn.com/2013/11/03/world/nuclear-energy-climate-change-scientists-letter/

Convery, Frank J. (2009) "Origins and development of the EU ETS," *Environmental & Resource Economics* **43**(3): 391–412.

Davenport, Coral and Alissa J. Rubin (2017) "Trump signs Executive Order unwinding Obama climate policies," *New York Times*. 28 March. Available at: https://www.nytimes.com/2017/03/28/climate/trump-executive-order-climate-change.html

Doyle, Alster and Janet Lawrence (2015) "Norway to complete $ billion payment to Brazil for protecting Amazon," *Reuters*. Available at: http://www.reuters.com/article/us-climatechange-amazon-norway-idUSKCN0RF1P520150915

EU (2013) *Climate Action: Building a World We Like, with a Climate We Like.* Publications of the European Union, Luxembourg.

European Commission (2015) Fact sheet: Questions and answers on the proposal to revise the EU emissions trading systems. Available at: http://europa.eu/rapid/press-release_MEMO-15-5352_en.htm

Findlay, Trevor (2010) "The Future of nuclear energy to 2030 and its implications for safety, security and non-proliferation: Overview," The Centre for International Governance Innovation. Available at: http://acuns.org/wp-content/uploads/2012/06/NuclearEnergyFuture.pdf

Gilbertson, Tamra and Oscar Reyes (2009) "Carbon trading: How it works and why it fails," *Critical Currents*. Dag Hammarskjold Foundation Occasional Paper Series. No. 7.

Green Fiscal Commission (2009) "The case for green fiscal reform: Final report of the UK Green Fiscal Commission," *Policy Studies Institute*.

Greenpeace (2009) "Slaughtering the Amazon". Available at: http://www.greenpeace.org/international/en/publications/reports/slaughtering-the-amazon/

Grolle, Johann (2013) "Cheap but imperfect: Can geoengineering slow climate change? Interview with David Keith," *Spiegel Online*. Available at: www.spiegel.de/international/world/scientist-david-keith-on-slowing-global-warming-with-geoengineering-a-934359.html

Hamilton, Clive (2014) "Geoengineering and the politics of science," *Bulletin of the Atomic Scientists* **70**(3): 17–26.

Harder, Amy, Colleen McCain Nelson and Rebecca Smith (2015) "Obama's new climate-change regulations to alter, challenge industry," *The Wall Street Journal*. Available at: https://www.wsj.com/articles/obamas-new-climate-change-regulations-to-alter-challenge-industry-1438560433

Hayward, Tim, "Why taking the climate challenge seriously means taking democracy more seriously". Available at: http://www.humansandnature.org/democracy-tim-hayward

Hillebrand, Rainer (2013) "Climate protection, energy security, and Germany's policy of ecological modernisation," *Environmental Politics* **22**(4): 664–682.

Himmelfreundpointer, Rainer (2012) "Dennis Meadows: There is nothing we can do," *Church and State*. Available at: http://churchandstate.org.uk/2013/04/dennis-meadows-there-is-nothing-that-we-can-do/

Huang, Yanzhong (2018) "Why China's good environmental policies have gone wrong," *New York Times*. 14 January. Available at: https://www.nytimes.com/2018/01/14/opinion/china-environmental-policies-wrong.html

Hulme, Mike (2014) *Can Science Fix Climate Change? A Case Against Climate Engineering*. Cambridge: Polity Press.

Jamieson, Dale (1992) "Ethics, public policy and global warming," *Science, Technology & Human Values* **17**(2): 139–153.

Jamieson, Dale (2014) *Reason in a Dark Time. Why the Struggle against Climate Change failed — and what it means for our Future*. New York: Oxford University Press.

Janicke, Martin (2004) "Industrial transformation between ecological modernisation and structural change". In: *Governance for Industrial Transformation*. (Eds.) Klaus, Jacob, Manfred Binder and Anna Wieczorek. Berlin: Environmental Policy Research Centre, pp. 201–207.

Keith, David (2013) *A Case for Climate Engineering*. Cambridge: MIT Press.

Kitagawa, Hideki (2017) "Environmental policy under President Xi Jinping leadership: The changing environmental norms". In: *Environmental Policy and Governance in China*. (Ed.) Kitagawa, Hideki. Tokyo: Springer Japan.

Leiserowitz, Anthony (2006) "Climate change risk perception and policy preferences: The role of affect, imagery and values," *Climatic Change* **77**: 45–72.

Li, Quan (2006) "Democracy and environmental degradation," *International Studies Quarterly* **50**: 935–956.

Lovelock, James (2009) *The Vanishing Face of Gaia*. New York: Basic Books.

Machin, Amanda (2013) *Negotiating Climate Change: Radical Democracy and the Illusion of Consensus*. London: Zed Books.

Machin, Amanda (2018) "Green democracy". In: *The Companion to Environmental Studies*. (Eds.) Castree, Noel, Mike Hulme and James Proctor. London & New York: Routledge, pp. 184–187.

Machin, Amanda (2019) "Changing the Story? The trajectory of the ecological modernisation discourse in the European Union". *Environmental Politics*. **28**(2): 208–227.

McKean, Cameron A. (2013) "Tokyo installing 300-ton pendulums in skyscrapers to keep them from falling down," *Next City*. Available at: https://nextcity.org/daily/entry/tokyo-installing-300-ton-pendulums-in-skyscrapers-to-keep-them-from-falling

Metcalf, Gibert (2009) "Designing a carbon tax to reduce US Greenhouse gas emissions," *Review of Environmental Ethics and Policy* **3**(1): 63–83.

Meyer, Robinson (2017) "The giant Trump climate order is here," *The Atlantic*. 28 March. Available at: https://www.theatlantic.com/science/archive/2017/03/trump-climate-eo/520986/

Morena, Camila, Daniel Speich Chassé and Lili Fuhr (2015) "Carbon Metrics Global abstractions and ecological epistemicide," *Heinrich Böll Foundation*. Available at: https://www.boell.de/sites/default/files/2015-11-09_carbon_metrics.pdf

Nature (2014) "Look ahead: Research into climate engineering must proceed — even if it turns out to be unnecessary," *Nature* **516**: 8. Available at: http://www.nature.com/news/look-ahead-1.16466

Nepstad, Daniel, *et al.* (2014) "Slowing Amazon deforestation through public policy and interventions in beef and soy supply chains," *Science* **344**: 1118–1123.

Neumayer, Eric (2002) "Do democracies exhibit stronger international environmental commitment? A cross-country analysis," *Journal of Peace Research* **39**(2): 139–164.

Nolte, Christoph, Arun Agrawala, Kirsten M. Silvius and Britaldo S. Soares-Filho (2013) "Governance regime and location influence avoided deforestation success of protected areas in the Brazilian Amazon," *PNAS* **110**(13): 4956–4961.

Nordhaus, William D. (2007) "To tax or not to tax: Alternative approaches to slowing global warming," *Review of Environmental Economics and Policy* **1**(1): 26–44.

Obama, Barack (2014) Speech to UN Climate Summit. Available at: http://www.whitehouse.gov/the-press-office/2014/09/23/remarks-president-un-climate-change-summit

Payne, Rodger (1995) "Freedom and the environment," *Journal of Democracy* **6**(3): 41–55

Pearce, David (1991) "The role of carbon taxes in adjusting to global warming," *The Economic Journal* **101**(47): 938–948.

Royal Society (2009) "Geoengineering the climate: Science, governance and uncertainty". Available at: https://royalsociety.org/policy/publications/2009/geoengineering-climate/

Runciman, David (2013), *The Confidence Trap. A History of Democracy in Crisis from World War I to the Present*. Princeton, New Jersey: Princeton University Press.

Schuppert, Fabian (2011) "Climate change mitigation and intergenerational justice," *Environmental Politics* **20**(3): 303–321.

Senit, Carole-Anne (2012) "France's missed rendezvous with carbon-energy taxation," *IDDRI Working Paper*. Available at: http://www.iddri.org/Publications/France-s-missed-rendezvous-with-carbon-energy-taxation

Stavins, Robert (2008) "Cap-and-trade or a carbon tax," *The Environmental Forum* **26**(5). Available at: www.hks.harvard.edu/fs/rstavins/Forum/Column_22.pdf

Stehr, Nico (2015) "Climate policy: Democracy is not an inconvenience," *Nature*. Available at: http://www.nature.com/news/climate-policy-democracy-is-not-an-inconvenience-1.18393

Stephan, Benjamin and Matthew Paterson (2012) "The politics of carbon markets: An introduction," *Environmental Politics* **21**(4): 545–562.

Stern, Nicholas (2006) *Stern Review: The Economics of Climate Change. Executive Summary*. Available at http://webarchive.nationalarchives.gov.uk/+/http://www.hm-treasury.gov.uk/stern_review_report.htm

Stern, Nicholas (2009) *Blueprint for a Safer Planet*. London: Vintage.

Stern, Nicolas, Isabella Neuweg and Patrick Curran (2017) "China's leadership on sustainable intrastructure: lessons for the world," *LSE Policy Brief*. Available at: http://www.lse.ac.uk/GranthamInstitute/wp-content/uploads/2017/07/Stern-et-al-2017_China-leadership_policy-brief.pdf

Szerszynski, Bronislaw and Maialen Galarraga (2013) "Geoengineering knowledge: Interdisciplinarity and the shaping of climate enginnering research," *Environment and Planning A* **45**: 2817–2824.

Szerszynski, Bronislaw and John Urry (2010) "Changing climates: Introduction," *Theory, Culture and Society* **27**(2–3): 1–8.

Tollefson, Jeff (2013) "Brazil's fight to save the Amazon and climate-change diplomacy," *Foreign Affairs*. Available at https://www.foreignaffairs.com/articles/brazil/2013-02-11/light-forest

Tollefson, Jeff (2015) "Stopping deforestation: Battle for the Amazon," *Nature* **527**(7545): 20–23. Available at http://www.nature.com/news/stopping-deforestation-battle-for-the-amazon-1.17223

UNEP (2012) The Emissions Gap Report 2012: A UNEP Synthesis Report. Available at http://unfccc.int/resource/docs/publications/tech_for_adaptation_06.pdf

UNFCCC (2017) Clean development mechanism. Available at: http://cdm.unfccc.int/index.html

van der Werf, *et al.* (2009) "CO$_2$ emissions from forest loss," *Nature Geoscience* **2**: 737–738.

Wara, Michael (2007) "Is the global carbon market working?" *Nature* **445**: 595–596.

Weiner, Sophie (2015) "Beat the drought: The walls of this subterranean house save water for you," *Fastcodesign*. Available at: http://www.fastcodesign. com/3046575/beat-the-drought-the-walls-of-this-subterranean-house-save-water-for-you

World Nuclear Association (2016) Radioactive Waste Management. Available at: http://www.world-nuclear.org/information-library/nuclear-fuel-cycle/ nuclear-wastes/radioactive-waste-management.aspx

Yale Forest Atlas (n.d.a), The Amazon Basin Forest. Available at: http:// globalforestatlas.yale.edu/region/amazon

Yale Forest Atlas (n.d.b), Forest Governance — Brazil. Available at: http:// globalforestatlas.yale.edu/amazon/forest-governance/brazil

York, Richard (2010) "The paradox at the heart of modernity: The carbon efficiency of the global economy," *International Journal of Sociology* **40**(2): 6–22.

Zeller, Tom (2013) "Carbon capture and storage: Global warming panacea, or fossil fuel pipe dream?" *Huffington Post*. Available at: http://www.huffingtonpost. com/2013/08/19/carbon-capture-and-storage_n_3745522.html

CHAPTER 7
Summary and Prospects

One of the messages of this book is that understanding the climate demands a multi-disciplinary perspective. We began, in Chapter One, by surveying the different understandings of climate. We show how climate has shifted from a scientific object that was seen as somehow external to society, to a complex and interconnected system that is partially affected by social structures, technological practices and energy cultures. Research into the climate system thus cannot only take place within the domain of natural sciences but must also involve the input of the social sciences and humanities. This message is of particular pertinence, we believe, at a time when climate science is under pressure, from both the general public and policymakers who need to decide if, and how, to respond to environmental risk and hazard, and from those who query the prominent role scientists are expected to play in policymaking.

Scientific climate research itself, as is the case for all research within both the natural and social sciences, is generated within a particular socio-historical context, which inevitably shapes all knowledge claims and the way they are put to work. We have thus attempted to contribute to a historical survey of such knowledge claims, attending to the scientific accounts of climate as embedded within, as well as contributing to, specific social structures.

In Chapter Two we consider the way in which scientific paradigms have changed over the centuries, not only as a result of the quest for greater understanding but also because of socio-economic pressures, political goals and cultural values. Returning to past paradigms and attending to previous scientific concepts is important because in the onset of advancing scientific research, the discoveries and ideas of only a few decades past can be quickly forgotten.[1] If scholars of the past are fortunate, their contributions may have been incorporated into contemporary scientific discourse by obliteration.[2] But mistakes and misconceptions are often not remembered, with the danger that they will be repeated with the same enthusiasm as in the past. To unpick not only the scientific "facts", but also the various interconnected social and political factors that fostered their emergence, and indeed, the conditions that may have encouraged their eventual forgetting, may make such repetition less likely.

This may well apply, for example, with the "scientific" claims of climate determinism. As we observe in Chapter Four, this highly problematic conviction that the climate is a determining factor of human civilisation and human behaviour, which has been prevalent in the past, may again be gaining prominence. This is precisely why examining the nineteenth and early twentieth century context for such claims might be a salutary tactic to avoid falling back into the same intellectual and political trap. After the painful experiences with climatic, racial and biological determinism of previous periods, it is perhaps understandable that social science research has effectively shied away from dealing with the impact of the natural environment on society. The consequence, however, is that for many years there was little in the way of "social science climate research", and thus the field of climate change was more or less exclusively monopolised

---

[1] While Svante Arrhenius is commonly recalled, only a few remember Guy Stewart Callendar, and Eduard Brückner is almost forgotten.

[2] A concept introduced by Robert K. Merton (1968: 27–29, 35–38) in *Social Theory and Social Structure*. In the process of the "obliteration by incorporation", both the original scholarly idea and the literal formulations of it are forgotten due to prolonged and widespread use and enter even into everyday use within the scientific community and everyday life, but no longer being linked to their originator; the idea or the discovery become similar to common knowledge in a field of discourse. We can reasonably assume that in the natural sciences the obliteration of the source of ideas, theories, concepts, methods, or findings by way of their incorporation in currently accepted knowledge, occurs much more regularly than is the case in social sciences where the history of ideas is in most disciplines a field of inquiry very much alive at any given time.

by the natural sciences. Social scientists who did engage in climate research tended to come from only a few social science disciplines; most predominantly economics (cf. Grundmann and Stehr (2010) and Clark and Yusoff (2017)).

This, we argue, constitutes a problematic oversight. For to recognise that climate does not constitute a total determinant is not to say that it has no impact whatsoever. Climate matters, but not as a *determining*, but rather as a more or less *conditioning*, influence. The extent to which a volatile climate may affect human individuals and societies over the coming decades and centuries is yet to be seen and may depend on the speed and scope of climate change, the tipping points that might be reached and feedback loops that might be set off. As we consider in Chapter Three, the impacts might be dramatic, but they are also unpredictable and unevenly distributed around the planet.

The sense that the world faces unprecedented and grave future dangers, risks and conflicts induced by anthropogenic climate change is mounting. In a study of the growing modern-day dilemmas, Pankaj Mishra (2017: 10) remarks: "The sense of the world spinning out of control is aggravated by the reality of climate change, which makes the planet itself seem under siege from ourselves." This sense of dramatic changes is exacerbated by the unprecedented speed of growth in scientific research and technological innovation. There is tentative hope for a scientific-technological mitigation of some aspects of the climate problem through new filtering techniques, alternative forms of energy and energy use, or improvements in energy efficiency (cf. Prins *et al.* (2013)). Others are less optimistic and warn that much more radical and structural change is needed to lessen the effects of a warming planet.

Given the retention time and accumulation of greenhouse gases in the atmosphere, and despite (or perhaps partially due to) efforts to reduce emissions, enhance resilience and implement new technologies, the relationship between climate and society is bound to change in new, differentiated and unpredictable ways. Climate change will surely involve significant changes in cultural values, political and economic processes, social structures and technological capacities. The possibility of drawing on previous experiences should not be undermined. As Wolfgang Behringer observes in his research on the Little Ice Age: "Even minor changes in the climate may result in huge social, political and religious convulsions." (Behringer, 2010: vii). It seems possible, then, to learn from responses, solutions and ideas coming from outside the dominant science and policy paradigms. And yet, paradoxically, the more that is known about the climate the less knowable it seems. What appears indubitable, however, is that responding to climate change will involve coping with ineradicable uncertainty alongside growing expertise and negotiating risk alongside innovation.

The way in which this negotiation occurs cannot be predicted, nor is there any clear "correct" path that social actors should take in responding to climate. Despite claims to the contrary, science cannot somehow override politics and force actors to compromise their positions on environmental issues (Ruser and Machin, 2017: 61). As we hope to have shown in Chapter Five, the information that science provides about the climate is valuable and complex and sophisticated, and it may encourage people to refine or revise their views, but it cannot ever produce certainty or predetermine political decisions that inevitably involve other sorts of values. When Philippe Sands QC then demands that an international court should scotch climate sceptics by ruling on the "facts" by "settling the *scientific* dispute" (Vaughan, 2015, italics added), does he not then risk a backlash against the technocratic suppression of political discussion over climate policy?

It may be no coincidence that, just as science is becoming increasingly important within policymaking, scientists are becoming increasingly subject to distrust (Machin *et al.*, 2017). Certainly, popular and political attitudes to both climate science and climate change are ambiguous: The once uncontested admiration, satisfaction and confidence in scientific knowledge and technical capacities have considerably waned and been replaced by a much more sceptical attitude (cf. Stehr (2004; 2006)). It is thus a challenge for scientists to put into practice the interdisciplinarity demanded in and of climate research. We need new types of science that include society as part of the earth's ecosystem, without reducing the non-rational, unpredictable, internal social dynamic to an environmental determinism.

Ultimately, human societies negotiate problems and solutions according to their own social, political, economic, technological and ecological dynamics. While international cooperation might be desirable, we might well doubt the chances of an agreement such as the voluntary Paris Accord of December 2015.[3] How realistic are its goals on emission reduction, adaptation and finance, especially after the withdrawal of the United States? While some believe that only a global solution can tackle a global problem, others doubt the wisdom of pursuing an overarching global strategy. The authors (Prins, 2010: 16) of the Hartwell Paper write:

> "Rather than being a discrete problem to be solved, climate change is better understood as a persistent condition that must be coped with and can only be partially managed more — or less — well. It is just one part of a larger complex of such conditions encompassing population, technology, wealth disparities, resource use, etc."[4]

In Chapter Six we consider some of the available options for policymakers and argue for a combination of approaches, and the enmeshment of climate within other policy areas and alongside other policy issues.

Many call for new forms of policymaking along with a more creative public discussion (Behringer, 2010: viii). Old assumptions that human society can be detached from its natural environment are coming under increasing pressure (Latour, 1991). The recent diagnosis of Anthropocene, the geological epoch in which human society has become a biogeophysical force capable of disrupting the entire earth-system, ruptures any presupposition that nature and society are distinct

---

[3] Paris Accord (2015) available on http://unfccc.int/paris_agreement/items/9485.php.

[4] See Goldthau (2017) on the need to *manage* the transition from fossil fuels with solar, wind, geothermal and biomass energy.

categories that can be easily separated. The words of Johann Wolfgang von Goethe written in 1825 seem prescient:

> "Above all we must remember that nothing exists or comes into being, lasts or passes, can be thought of as entirely isolated, entirely unadulterated. One thing is always permeated, accompanied, covered, or enveloped by another; it produces effects and endures them. And when so many things work through one another, where are we to find the insight to discover what governs and what serves, what leads the way and what follows?" (Goethe, cited in Tantillo (2002: 140)).

In debating, researching and understanding climate and climate change we would do well to heed not only the interconnections of the climate system, but also the processes, practices and tensions through which science, society, nature and climate permeate, accompany, cover and envelop each other.

References

Behringer, Wolfgang (2010) *A Cultural History of Climate*. Cambridge, Malden: Polity Press.

Clark, Nigel and Kathryn Yusoff (2017) "Geosocial formations and the anthropocene," *Theory Culture & Society* **34**: 3–23.

Goldthau, Andreas (2017) "The G20 must govern the shift to low-carbon energy," *Nature* **546**: 203–205.

Grundmann, Reiner and Nico Stehr (2010) "Climate change: What role for sociology? A response to Constance Lever-Tracy," *Current Sociology* **58**: 897–910.

Latour, Bruno (1991) *We Have Never Been Modern*. Harvard University Press.

Machin, Amanda, Alexander Ruser and Nikolaus von Andrian Werburg (2017) "The climate of post-truth populism. Science vs. the people," *Public Seminar*. Available at http://www.publicseminar.org/2017/06/the-climate-of-post-truth-populism/

Merton, Robert K. (1968) [1949, 1st ed.] *Social Theory and Social Structure*. New York: Free Press.

Mishra, Pankaj (2017) *Age of Anger. A History of the Present*. London: Allen Lane.

Prins, Gwythian, *et al.* (2013) *The Vital Spark*: *Innovating Clean and Affordable Energy for All*. London: LSE Academic Publishing.

Prins, Gwythian, *et al.* (2010) *The Hartwell Paper. A New Direction for Climate Policy after the Crash of 2009*. London: London School of Economics.

Ruser, Alex and Amanda Machin (2017) *Against Compromise: Sustaining Democratic Debate*. London: Routledge.

Stehr, Nico (2004) *The Governance of Knowledge*. New Brunswick: New Jersey: Transaction Books.

Stehr, Nico (2006) *Knowledge Politics. Governing and Consequences of Science and Technology*. New York: Routledge.

Tantillo, Astrida Orle (2002) *The Will to Create: Goethe's Philosophy of Nature*. Pittsburgh: University of Pittsburgh Press.

Vaughan, Adam (2015) "World court should rule on climate science to quash sceptics, says Philippe Sands," *The Guardian*. 18 September. Available at: www.theguardian.com/environment/2015/sep/18/world-court-should-rule-on-climate-science-quash-sceptics-philippe-sands

Author Index

Subject Index

www.ingramcontent.com/pod-product-compliance
Lightning Source LLC
Chambersburg PA
CBHW050601190326
41458CB00007B/2127